Advances in Cyber Security

Advances in Cyber Security

Technology, Operations, and Experiences

Edited by D. Frank Hsu and Dorothy Marinucci

FORDHAM UNIVERSITY PRESS

NEW YORK 2013

Library of Congress Cataloging-in-Publication Data

Advances in cyber security : technology, operations, and
experiences / edited by D. Frank Hsu and Dorothy Marinucci. —
First edition.
        pages cm
    Includes index.
    ISBN 978-0-8232-4456-0 (cloth : alk. paper) —
ISBN 978-0-8232-4457-7 (pbk. : alk. paper)
    1. Internet—Security measures.   2. Cyberspace—Security
measures.   3. Computer networks—Security measures.
I. Hsu, D. Frank (Derbiau Frank), 1948–
    TK5105.875.I57A343 2013
    005.8—dc23

                                        2012039718

Printed in the United States of America

15 14 13   5 4 3 2 1

First edition

CONTENTS

Partners in Cybercrime | *Eileen Monsma, Vincent Buskens,*
Melvin Soudijn, and Paul Nieuwbeerta

Securing IT Networks Incorporating Medical Devices | *Nicholas J.*
Mankovich

Information Technology for a Safe and Secure Society
in Japan | *Kazuo Takaragi*

Preface

The advent of network and digital information technologies has ushered in a myriad of new and diverse opportunities in our lives and at our jobs. Cutting-edge technologies have transformed the information society into a vibrant cyber-physical-natural ecosystem. However, these same technologies are being used by adversaries in a variety of attacks and exploitations that disrupt the cyberspace ecosystem's balance.

Recognizing the emerging field of cyber security as an international response to the challenge posed by cyber threats to individual well-being, organizational foundation, multinational peace, and global enterprise, Fordham University and the Federal Bureau of Investigation, New York Division, organized the first International Conference on Cyber Security (ICCS) in 2009 (ICCS 2009) and held the second ICCS in 2010. ICCS brought together cyber security leaders from government, industry, and academia who design, develop, manage, and sustain various aspects of secure cyberspace ecosystems.

Advances in Cyber Security: Technology, Operations, and Experiences brings together their thoughts, analyzes key issues confronting cyber security experts, and offers global solutions to emerging issues and problems. It is designed to give cyber citizens and cyber professionals increased knowledge in the areas of cyber technology, cyber operations, real-life cyber security experiences, and sensible cyber security policy.

Advances in Cyber Security begins with an expository chapter by D. Frank Hsu that presents an overview of the cyber world and explains how the digital landscape evolved and now operates. Hsu's broad thesis of how the cyber, physical, and natural worlds are interrelated shows how the cyberspace ecosystem is interconnected, and how these three elements interact with each other. He then conceives feasible and practical solutions for defending the cyberspace ecosystem against threats from cyber attacks and cyber exploitation.

Following this overview, *Advances in Cyber Security* looks at current thinking about cyber security technology with chapters by leading cyber security researchers: Ruby B. Lee from Princeton University, Paul Syverson with the Naval Research Labs, Hira Agrawal, Tom Bowen, and Sanjai Narain from Applied Communication Sciences (formerly Telcordia Technologies), and Akio Sugeno from Telehouse America.

How the technological breakthroughs presented by these authors can be utilized is discussed in the next section on operations, in chapters by Andrew Lewman of the Tor Project, Kuan-Tsae Huang and Hwai-Jan Wu of Taskco Corporation, Adam Palmer from Symatec Corporation, and Eileen Monsma, Vincent Buskens, Melvin Soudijn, and Paul Nieubeerta, all with the National High Tech Crime Unit (NHTCU) of the Netherlands.

Both technology and operational theory make up the topics covered in the third section, which deals in real-life experiences with contributions from Nicholas J. Mankovich from Philips Healthcare Product Security and Privacy, Kevin Kelly from Fordham University's faculty, Edward M. Stroz of Stroz-Freidberg, LLC, and Kazuo Takaragi of Hitachi Corporation.

Finally, *Advances in Cyber Security* addresses the question of partnership and policy decisions with the key people involved in the everyday enforcement of cyber protection. InfraGard's views are presented by National Board Chairperson Kathleen L. Kiernan and President Dyann Bradbury. Preet Bharara, United States District Attorney for the Southern District of

New York, Howard A. Schmidt, special assistant to the president of the United States and federal cyber security coordinator, and Robert S. Mueller III, director of the Federal Bureau of Investigation, present their views of what must be done to stop the increasing cyber threats from disrupting people's lives.

This book is a genuine collaboration of people from government, academia, and industry who are committed to a peaceful planet. We thank the authors, who have contributed time and effort to writing chapters for this volume. We certainly hope this book will serve as a catalyst for a series of future dialogues on the most important and emerging issues in building a secure and safe cyberspace ecosystem. Thanks to the faculty and students of Fordham University's computer and information science department for their technical support, and to the student ambassadors who assisted at the conference. The editors would like to acknowledge Father Joseph M. McShane, S.J., president of Fordham University, Maureen Topf from Fordham University, and George Venizelos, Austin Berglas, and Anthony Ferrante from the FBI for their vision and support in making ICCS possible. The enthusiasm and passion of so many citizens who bonded together at ICCS is a great testimony to their passion for ensuring a secure and sustainable cyberspace ecosystem in the future.

We want to express our thanks to Fordham University Press director, Fredric Nachbaur, Fordham University Press managing editor, Eric Newman, and Pamela Nelson from Westchester Publishing Services for their support in making this publication possible.

Advances in Cyber Security

Building a Secure and Sustainable Cyberspace Ecosystem: An Overview

D. Frank Hsu

This overview provides a historical and contemporary perspective on various issues pertaining to the security and sustainability of the emerging cyberspace, which is embedded with intelligent networking sensors and systems, as well as information technology appliances and services. In particular, it explores how to build a secure and sustainable cyberspace ecosystem in the combined cyber-physical-natural (CPN) world. Its three sections give an overview of the emerging interconnected complex cyberspace, review the infrastructure for the combined CPN cyberspace, and provide a list of intellectual tools for collaboration, education, and partnership in order to build and sustain a secure cyberspace ecosystem. Finally, this overview emphasizes the need for establishing an international consortium and a cyber security informatics framework to facilitate coordination and collaboration among cyber security professionals and practitioners from government, academia, and industry across national and international boundaries. I hope this overview will serve as a catalyst for all cyberspace stakeholders to develop new ideas and construct novel solutions to address global challenges and create opportunities for a secure cyberspace ecosystem.

The Emerging Complex Cyberspace

This section provides a holistic overview of the emerging complex cyberspace in terms of four diverse perspectives: digital landscape, technology capacity, the fifth domain, and the fourth paradigm.

THE DIGITAL LANDSCAPE

Human civilization has evolved from an agricultural society to an industrial society and then to an information society. Since the Internet's establishment in the 1960s and the World Wide Web's conception in the 1990s, the number of information users and providers has increased exponentially. A vast amount of information content with great variety has been created, stored, processed, transmitted, and utilized. These range from simple text, speech, and images, to histories of interactions with friends, colleagues, sources, sensors, systems, and proxies [85]. Raw data sources on the Internet also include sensor readings from GPS, GIS, and RFID devices, medical devices such as MRI and EEG, and other embedded sensors, systems, and robots in one environment or in the infrastructure of the information society.

Communication conduits for transmitting digital content have evolved in speed and bandwidth from twisted pairs for voice to coaxial cables for images and finally to optical fibers for voice, text, and images. The networks that facilitate this rapid communication use wireline, wireless, satellite, and Internet interconnections. At the same time, the point of contact—the device we use to access information or information network—improves in size, scope, and scale. These devices, which we now call *information appliances*, include radio, TV, iPods, iPhones, BlackBerrys, laptops, notebooks, desktops, and iPads. The capabilities of these devices increase, yet their cost decreases. These devices constitute a host of pervasive computing and communication systems in the cyber world.

In the physical and natural world, human beings have made great strides toward personalized medicine and public health due to the scientific revolution in genetics, genomics, and proteomics in the past century, but more so since the 1950s. In addition, a renewed emphasis on translational science (from bench to bed side) has enhanced capability in the diagnosis, screening, and treatment of diseases and disorders. More recently, the concept of a molecular network, which connects molecular biology to clinical medicine through the omics pathways (metabolomics, genomics, proteomics, and transcriptomics) has become a foundation and a major focus of translational science [93].

On the other hand, due to the advent of imaging technology, cellular and molecular biology, bioinformatics, systems and computational neuroscience, and brain and cognitive informatics, we have increased our understanding of the structure and function of the neural system. Brain connectivity and its computational power has enabled us to know more about basic sensory and motor systems, memory, perception, cognition, and even behavior than we did before. Any change in human physiology resulting from the function of the neuronal network in the brain, with 100 billion neurons, would lead to a change in human cognition and behavior [7, 89].

THE TECHNOLOGY CAPACITY

Networking, information, and communication technologies (NICT) are the driving engines for the information society, where economic wealth is generated, political power is exercised, and cultural codes are created [15]. It would be helpful and interesting to quantify the amount of information processed by the information society. Lesk [71] asked, "How much information is there in the world?" Pool [91] estimated the growing trends of information (measured as amounts of words) transmitted by seventeen major communications media sources from 1960 to 1977 in the United States. Lesk's study in 1977 was the first empirical undertaking to show the declining relevance of print media and the growing prevalence of electronic media. For digital storage, Gantz et al. [42] estimated that all the empty or usable space on hand devices, tapes, CDs, DVDs, and volatile and non-volatile memory in the market in 2007 equaled 264 exabytes (1 exabyte = 10^{18} bytes).

Hilbert and Lopez [54] went one step further to study the amount of information actually handled by the hardware, rather than merely stored in the hardware itself. Data contained in storage and communication hardware capacity were converted into information bits by normalizing the compression rates. This highlights the fact that information resources have different degrees of redundancy, where the redundancy is primarily determined by the content in question, such as text, images, audio, or video [29, 96]. Communication capacity can be divided into two groups: one provides unidirectional downstream capacity to diffuse information (i.e., broadcasting) and the other provides bidirectional upstream and downstream channels (i.e., telecommunications). Therefore, "computation" can be viewed as the repeated transmission of information through time (communication) and space (storage), guided by an algorithmic procedure [107].

By tracking 60 analog and digital technologies during the period from 1986 to 2007, Hilbert and Lopez [54] were able to estimate the world's technological capacity to store, compute and communicate information. In 2007, humankind was able to store 2.9×10^{20} optimally compressed bytes, an increase of 23 percent in globally stored information. Likewise, humankind was able to communicate almost 2×10^{21} bytes, with the capacity increased for bidirectional telecommunications at 28 percent per year and for unidirectional information diffusion through broadcasting channels at a modest growth rate of 6 percent. As for the use of general-purpose computing capacity, humankind was able to carry out 6.4×10^{18} instructions per second in 2007. It grew at an annual rate of 58 percent. This number of instructions per second is similar to the maximum number of neural impulses executed by one human brain per second.

On the other hand, the 2.4×10^{21} bits stored in all technological devices in 2007 are approaching the roughly 10^{23} bits stored in the DNA of a human adult. However, this number is still relatively small compared to the 10^{90} bits stored in the observed universe [54, 75].

In summary, cyberspace has gone digital. With the advent of networking and optical fiber technology, technology capacity depends not only on the structure but also the function of the interconnection network. Telecommunication has been dominated by digital technology since 1990 (99.9 percent of telecommunications were in digital format in 2007), and the majority of technological memory has also been in digital format since the early 2000s (94 percent digital in 2007) [54].

THE FIFTH DOMAIN

Computing, information, and communication technologies (CICT) have transformed economies and given countries advantages in defense, global competitiveness, and other critical infrastructures. For example, CICTs also provide the ability to send remotely piloted aircraft across the world to collect information and identify targets. However, the spread of digital technology comes at a cost: it exposes the vulnerability of individuals, organizations, societies, and nations to digital attacks. After land, sea, air, and space, cyberspace has become the "fifth domain" of defense and warfare [25, 32, 109]. Most vulnerable to attack are infrastructure components that depend on digital technologies. In his book, Clarke [26] envisages a scenario where a catastrophic breakdown can happen in less than a few minutes. Military and civilian e-mail and other communication systems are brought to a complete halt, air-traffic-control systems cease to function, freight and metro trains derail, the electrical power grid is interrupted, and oil refineries and gas pipelines explode. These events are followed by a shortage of food supplies and a financial crisis, leading to more systemic failure in the interconnected and instrumented society [109].

Cyberspace, and its digital infrastructure, was declared a "strategic national asset" by President Barack Obama when he appointed Howard Schmidt as his National Cybersecurity Coordinator in December 2009. In May of 2010, the Pentagon established the new U.S. Cyber Command (CYBERCOM) headed by General Keith Alexander, Director of the National Security Agency (NSA). CYBERCOM was tasked to protect the military ".mil" domain and support the government ".gov" and the commercial ".com" domains. Other countries such as Great Britain, China, Russia, and Israel are also organizing infrastructure for cyber defense and prevention of cyber attacks and cyber exploitations [109].

THE FOURTH PARADIGM

Scientific inquiry and knowledge discovery has a long history and has gone through at least three different phases. According to Jim Gray [53], a thousand years ago, science was rather *empirical* and it was used to describe natural phenomena. Scientists used their five senses and power of critical reasoning to measure, describe, and establish theories of the physical world. In the last few hundred years, in particular, since the 1600s, *theoretical* branches of science used models, methods, hypotheses, and generalizations to explain their findings. Then, in the last few decades, scientists have increasingly used *computational* models and simulations to investigate complex phenomena. More recently, due to the advent of imaging and sensor technologies and modalities, large and diverse data have been generated and collected. These include structured and unstructured data from a variety of domain areas related to climate change, ecological systems, global health, biomedicine, health care, neuroimaging, astronomy, social networks, and cyber security. Most disciplines in the natural sciences, engineering, social sciences, humanities, and professional studies are finding the data deluge to be extremely challenging [34].

Tremendous opportunities can be realized if we know how to manage large amounts of data, including their collection, curation, access, simulation, and visualization. Today's science is moving toward exploration of diverse data using a multifaceted, empirical approach, theoretical model, and computational simulation, unified to curate and analyze databases and other electronic file systems. Mathematical, computational, and statistical methods are also used to perform data mining, machine learning, information fusion, and knowledge discovery. This latest "fourth paradigm" is what Jim Gray called *e-science* [53]. He also described this as the beginning of a host of interdisciplinary fields called computational-*x* and *x*-informatics, where *x* stands for any discipline. For example, if *x* is biology, then we have computational biology and bioinformatics. Similarly, we have computational neuroscience and neuroinformatics. We also have computational social science and socioinformatics. In the context of cyber security, we have computational cyber security and cyber security informatics.

A Cyber-Physical-Natural Infrastructure

This section reviews the combined cyber-physical-natural cyberspace infrastructure, including ubiquitous networks, internetworking, internet protocol addressing, IPv6, pervasive systems, and social interaction networks.

UBIQUITOUS NETWORKS

All computing systems and devices, communication routers and switches, and information appliances and grids are interconnected via local networks, metropolitan networks, and wide area networks. The Internet is really a network of interconnected networks, which consists of different layers with interfaces between layers at each host (or each node). The Open System Interconnection (OSI) reference model, suggested by the International Organization for Standardization (ISO), consists of seven layers—the physical, data link, network, transport, session, presentation, and application layers [33, 101].

The three lower layers (physical, data link, and network) of the OSI model function closely within the communication subnet boundary [101]. The three layers act together as an interconnection from host to router and back to router on another, distant host machine. The physical layer is concerned with raw lists over a communication channel, while the data link layer is concerned with transmitting and acknowledging data frames (normally a few hundred or a few thousand bytes). The network layer is more concerned with routing *packets* (a combination of frames) from source to destination and controlling the operation of the subnet. These three layers at host node and destination node communicate with each other using the Internet subnet protocol and the host-router protocol at their respective layers.

The network layer as described by the OSI reference model uses packet switching instead of the circuit switching used by voice networks in telephone communication systems. Packet switching can remove the artificial constraints imposed by splitting links into circuits. However, this flexibility comes with risks. In the Internet, with packet switching we have no control over congestion, which circuit switching provides. This led to the Internet congestion collapse in 1988. Since then, the Internet community has been using Van Jacobson's Transmission Control Protocol (TCP) in which the congestion is controlled by the end systems on their own [62].

Due to the recent proliferation of multihomed mobile devices, data centers, and cloud computing systems, multiple path transmission protocols at the transport layers have become a hot topic in the network systems community. Multipath transmission allows users to send packets along different paths. According to Wischik [116], one obvious benefit of multipath transmission is greater reliability. One example of this enhanced reliability is the capability for one's phone to use WiFi when it can be switched to a cellular network if needed. Another benefit of multipath transmission is an extra degree of freedom and flexibility in sharing the networks. Although multipath transmission has the advantage of removing the artificial constraints imposed by "splitting" the network's total capacity into separate links, it brings the disadvantage of losing congestion control. Key et al. [65] investigate the benefits that accrue from the use of multiple paths by a session, coupled with the rate of control over these paths. In particular, they study data transfers under both coordinated and uncoordinated control where the rates over the paths are determined as a function of all paths and independently over each path, respectively.

In summary, Key et al. [65] suggest, once the crude control of "each flow may use only one path" is no longer enforced, that some new control be put in place which can be achieved by end systems on their own. In comparison, if multipath is considered as packet switching 2.0, then multiple congestion control is needed as TCP 2.0. In the future, multipath routing, communication, and congestion control will be tested on a larger scale and with wider scope before it can be implemented. A set of disjoint paths between a pair of two nodes is called a *container*. The width and length of a container in graphs, groups, and networks have been studied extensively since 1994 [57].

INTERNETWORKING

Many general information users and providers assume (or perceive) the Internet to be a single homogeneous network, with each machine using the same protocol in each layer and the same application at each node. However, when we examine the Internet more closely, it becomes apparent that it is a conglomerate of thousands of networks with each node functioning at all seven layers. At the network layer, the Internet can be considered as a group of subnetworks or *Autonomous Systems* (ASes) that are connected by routers, switches, and optical fibers. Although there is a similarity to routing algorithms between intra-subnet routing and inter-subnet routing, differences do exist because of the two-tier architectural structures and the multi-protocol routers [101].

The two-level routing algorithm consists of an interior gateway protocol used within each network (AS), and an exterior gateway protocol used between the networks (ASes). Since each AS is operated by a different organization, an Internet service provider (ISP) or a network service provider (NSP), it can use its own routing algorithm inside the system: an interior gateway protocol. In 1988, the Internet Engineering Task Force (IETF) began its work on a new generation interior gateway protocol. As a result, the routing protocol OSPF (Open Shortest Path First) became a standard in 1990 [101]. Between different ASes, an exterior gateway protocol called Border Gateway Protocol (BGP) is used to control routing protocols and policies to be enforced in the inter-AS traffic (see the description of BGP in [101]). These policies involve security, political, cultural, or economic considerations which are manually configured into each BGP router, in addition to the protocol itself [101].

The communication protocols we have mentioned are for IP communication between one sender and one receiver on a fixed physical connection. For some applications, it is desirable to be able to send to a fairly large number of destinations simultaneously. These tasks include updating replicated or distributed databases, transmitting stock quotes to multiple brokers, sending e-mails to all employees in a company, and executing digital conference calls. The IP communication protocol does support these functions using a multicasting algorithm called Internet Group Management Protocol (IGMP) [101]. In addition, many users of the Internet use portable computers and many more have the need to access the World Wide Web or their e-mails when they are traveling. Mobile IP issues and solutions were described and proposed in 1995 [63]. Wireless access from computers, cellular phones, and other smart devices has become popular now.

INTERNET PROTOCOL ADDRESS AND IPV6

The design of TCP/IP (Transmission Control Protocol/Internet Protocol) started in 1973 by Vinton G. Cerf and Robert Kahn [17, 18]. It was finally concluded that TCP should be split into two protocols: a simple Internet Protocol that would carry a datagram end-to-end through packet networks and interconnected gateways, and a Transmission Control Protocol that processed the flow and sequencing of packets exchanged between hosts on the Internet [73]. This split enabled the real-time and nonsequenced packet delivery to support various forms of packet transmissions such as voice, video, radar, and other real-time streams. It was also decided that 32 bits of address space were needed to address the Internet with 8 bits of network and 24 bits of host. It was called IPv4 because it was reached in the fourth iteration of its research and development. This address space of 32 bits was able to accommodate over 2.1 million addresses. The assignment of address space was allocated by the Internet Corporation for Assigned Names and Numbers (ICANN), formerly the Internet Assigned Number Authority (IANA).

Early in 2011, ICANN announced its allocation of the last five of these large 24-bit chunks of address space [73]. Back in the early 1990s, the Internet Engineering Task Force (IETF) extended the address format for the Internet called IPv6 (IPv5 was about stream applications and was abandoned due to its inability to scale). IPv6 included a small number of new features and a format intended to expedite processing. However, the principal advantage of IPv6 was its address space of 128 bits for each of the source and destination hosts. As such, it could cover a multitude of many trillions of trillions of addresses. In the near future, we will see the migration of 32-bit IP addresses (IPv4) to the 128-bit address format (IPv6). A map of the IPv6 Internet was created by Lumeta Corporation [27, 73].

PERVASIVE SYSTEMS AND SOCIAL NETWORKS

As stated earlier in the section "The Fourth Paradigm," large data sets are collected by sensors and imaging technologies from a variety of sources and domains. These sensors are pervasive in the physical world, as the following three examples illustrate. RFIDs (radio-frequency identification devices) are used to track a product shipped from its manufacturer to a wholesale outlet or retail shop. The shipment can be domestic or international, and either air freight or ocean cargo over a distance of thousands of miles. The second example is a wireless sensor network that can monitor active and hazardous volcanoes [115]. This is a good example of using sensor networks as a data collection tool to enhance the scientific discovery process. The third example is really down to earth [41]. In order to make better wine, wineries such as Napa Valley's Palmaz Winery and Vineyard

29 (both make Napa cabernets), Clos de la Tech (which makes pinot noirs), Rhys Vineyards (which makes pinot noirs and chardonnays in the Santa Cruz mountains), and Sonoma County's Scherrer Winery (which makes pinots) use sensors to measure grapes' Brix (ripeness), total acidity, and pH levels. This information is used to help determine the best time to harvest, in addition to the traditional cues based on the look and the taste of the fruit.

John McCarthy, a computer science and artificial intelligence pioneer, predicted in 1961 that computing may someday be organized as a "public utility just as the telephone system on the electricity power grid." The evolution of complexity has moved toward greater specialization and distribution of resources [5]. Cloud computing, as one of the latest computing paradigms, aims to deliver application services over the Internet and coordinate hardware and systems software in the data centers that provide these services [6]. Cloud computing has several advantages such as convenience and transparency, dynamic elasticity and scalability, and economy. However, these benefits come with important computing issues including security and privacy. Some researchers described a virtual private storage service which provides the security of a private cloud and the cost savings of a public cloud [5]. Just as large ISPs use multiple network service providers so that failure by a single company will not take them off the air, Armbrust et al. [6] propose to use multiple cloud computing providers in order to provide plausible solutions to very high availability and robustness.

In the natural world, people started to examine how much they are connected to each other and calculate the degrees of separation between two persons in the population. The calculation was done with a test called Lancszemeck (or chains) in 1929 by Frigyes Karinthy or a game called Messages in 1961 by Jane Jacobs [9, 61]. According to Barabasi [9], Karinthy's 1929 insight that people are linked by less than or equal to five links was the first published concept of the modern phenomona "six degrees of separation." The experiment conducted by Stanley Milgram in the 1960s aimed to find the number of intermediate persons letters have to travel through if they are sent to Boston, Massachusetts from Omaha, Nebraska [81]. Milgram found that the median number of intermediate persons was 5.5, which was rounded to 6. The phrase "six degrees of separation" was in fact used the first time by John Guare in his 1991 Broadway play of the same title [52]. These are just a few early examples of attempts to understand the structure of our social network. Recent examples include the collaboration network (such as the Erdös number [35, 49]), actors' networks [40], and the popular Facebook and Twitter. Other pervasive systems exist within various interconnection networks including power grid, transportation network, food web, network of neurons, and the World Wide Web (WWW) [22, 66, 110, 111, 112]. All these interaction networks, and the WWW in particular, have provided a vibrant marketplace, which we call the *information marketplace*, for a variety of information exchanges and transactions.

Building a Secure Cyberspace Ecosystem

This section explores how to build and sustain a secure cyberspace ecosystem. These include immersion of a cyber-physical-natural cyberspace ecosystem; cybersecurity informatics; collaboration among government, academia, and industry; and education and sustainability.

IMMERSION OF A CYBER-PHYSICAL-NATURAL (CPN) ECOSYSTEM

The 2009 DARPA Red Balloon Challenge explored how the Internet and social networking can be used to solve a distributed, time-critical, geo-location problem. Participating teams had to find 10 red weather balloons deployed at undisclosed locations throughout the continental United States. The first team to find all ten locations would receive a $40,000 prize [102]. The 2009 challenge also marked the fortieth anniversary of the first remote log-in to the ARPANET (October 29, 1969), an event known as the birth of the Internet. Since the 1990s, the World Wide Web has become one of the most (if not the most) powerful medium for sharing, creating, and distributing information in history. Although there are some who are concerned about the future of the World Wide Web, the use of the Web has rapidly spread to many other areas in social, government, and business environments [80]. These include everyday artifacts, online banking and stock trading, official government documents, banking and financial services, book and electronics retailing, and music and video purchasing.

More recently, the Web has begun to transform the political landscape across the world [12, 94]. In the health area, efforts to mine electronic health records provide ample opportunities for improving the delivery, efficiency, and effectiveness of health care [92]. These capabilities have been also fueled by the advent of integrative technologies such as implementing dual-core processors to improve smart phone performance [87], installing brain-computer interfaces so that the brain's electrical signal (in the neural circuits) would enable people without muscle control ability to physically interact with the outside world [79], and enacting the network neutrality principle by which the owners of broadband networks or network operators with BGP capability cannot discriminate against any traffic transmitted through them [76]. The immersion of the cyber world and the physical world has been realized rapidly. It has become more difficult to make the distinctions between human and computer intelligence and between the virtual (cyber) and the real (physical) world [50]. Researchers around the world (cyber or physical world or both) gathered in March 2011 to discuss and map the boundaries between the two emerging fields of Web science and network science [118].

Crowdsourcing systems aim to enlist a multitude of people (or cyber citizens) to help solve a wide variety of problems using the Internet and the Web [38]. Notable among those crowdsourcing systems are examples such as Wikipedia, Linux, Yahoo! Answers, and mechanical task-based systems. Crowdsourcing can be viewed as an intriguing form of computing that uses market data and the prediction markets (PM) method to aggregate a large amount of information from various individuals (or cyber citizens) to generate a forecast of the market [47, 60]. While we can infer social networks, social media, online advertising, and social games from physical interactions, each of us might want to have a personal social media strategy [39, 43, 44, 46, 67, 98].

Figure 1 summarizes and illustrates the immersion (form and function) of the combined CPN Cyberspace Ecosystem. In this combined world, the seven layers of the OSI model for physical networks, mentioned earlier in the "Ubiquitous Networks" section, are simplified to three—Structure Layer, Network Layer, and Application Layer—to match with the three layers of the combined CPN world—Computing Utilities, Interconnection Network, and Information Marketplace, respectively. (See the "Ubiquitous

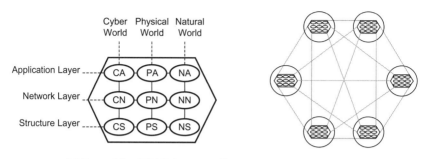

(a) Form of the CPN Cyberspace Ecosystem

	Cyber world (Technology)	Physical world (Operation)	Natural world (Experience)	Combined CPN Cyberspace
Application Layer (Information Marketplace)	e-mail, search, entertainment, cloud computing, e-business, e-commerce, ...	Work, life, research, entertainment, ...	Personalized medicine, translational science, environmental monitoring, ...	Social network, crowd sourcing, online auctions, cloud computing, ...
Network Layer (Interconnection Network)	Internet, www, Intranet, ...	Power grid, telecommunications networks, highway systems, ...	Molecular networks, neural circuits, brain connectivity, food web, ...	Ubiquitous network, multimedia network, autonomous systems, ...
Structure Layer (Computing Utility)	Bits, bytes, frames, packets, routers, switches, computers, ...	Individual organization, society, nation, state, ...	DNA, genes, proteins, patients, ...	Pervasive systems, information appliances, ...

(b) Function of the Combined CPN Ecosystem

Figure 1. Immersion of the Cyber-Physical-Natural (CPN) ecosystem.

Networks," "Internetworking," and "Pervasive Systems and Social Networks" sections earlier in this overview.)

In September 2010, four network experts were invited by *Discover Magazine* and the National Science Foundation to a discussion "Weaving a new web and the future of life online" [113]. Several ideas mentioned were: IPv6 implementation, cloud computing, surgery over the Internet, next generation Internet and Web that can make seamless transitions between cellular and WiFi networks, network centrality principle, message ferry as a dynamic Internet node, and crowdsourcing. Since the combined CPN ecosystem is an interconnected and complicated system, cyber threats (in the fifth domain) have to be dealt with by a combination of efforts including emerging new technologies on computing utility, operation practices in the interconnected network community, and real world experiences in the information marketplace. Each of the three layers in the CPN ecosystem hierarchy applied to each of the three CPN worlds in cyberspace would create at least nine interconnected and interleaved subdomains and related tasks in order to provide sound global and sustainable solutions for possible cyber security threats and cyber exploitation (refer to Figure 1).

In the combined CPN cyberspace ecosystem, (computing utility) systems, (interconnection network) software, and (information marketplace) services are all interrelated. A problem at any node in the complex cyberspace network—being attacked or exploited by adversaries—can have a cascading and systemic effect on the whole ecosystem. These systems range from the sensors embedded in Napa Valley's vineyard to the servers installed at the cloud computing data center in Hoboken, New Jersey, or from the switches and routers at the peering center among autonomous systems to the services on a social networking site using analytical methods and tools. Liu, Slotine, and Barabasi [74] study the controllability of thirty-seven complex networks such as those in the cyber world, physical world, and natural world and identify *driver nodes* that can control the entire dynamic complex system using a time-dependent control mechanism. It was shown that gene regulatory networks in the natural world, social networks in the cyber world, and engineered networks in the physical world (such as power grids or electronic circuits) have 80 percent, 20 percent, and less than 20 percent, respectively, as driver nodes. The network is more reliable and immune to hostile outside attacks if it needs higher numbers of driver nodes. Networks in the physical world such as the power grid or physical layer of the Internet are easier to control. But it will depend on who is trying to control it in the natural world. Since the cyberspace ecosystem is the function of a combination of three distinctive worlds, an optimal strategy to secure the ecosystem should harness the structural and functional capabilities among and leverage the diversity between individual network systems in all those three different worlds.

CYBERSECURITY INFORMATICS (CSI)

In the twenty-first century, the cyberspace ecosystem is covered with computing devices, communication gadgets, and information appliances. The physical world is instrumented with sensors and devices and interconnected by networks. In addition, the living systems of the natural world operate as functions of genes and proteins interconnected by molecular networks [92]. Technological capability in data collection and curation has improved so much that we have to explore efficient and effective ways of turning data into knowledge as described in the fourth paradigm [53]. For example, once the Large Synoptic Survey Telescope (LSST), a next-generation state-of-the art survey telescope, goes online in 2016, it will be capable of surveying the entire sky in only three nights [45]. Today's data deluge enables us to develop new approaches to visualize, catalog, and analyze enormous data sets.

As Jim Gray pointed out [53], the conversion from data to knowledge is a complex process of the pipeline:

Data   →   Information   →   Knowledge

where computational-x and x-informatics are needed for each application domain x or each world, be it cyber, physical, or natural. The consensus is to have a data-centric, evidence-based approach to solving problems and making decisions for the CPN cyberspace ecosystem.

Informatics is an emerging discipline consisting of methods of acquiring and seeing patterns in large diverse data sets, processes of information integration, and applications of extracting useful knowledge from information for making decisions and taking actions [51]. Hsu, Kristal, and Schweikert [55] point out a critical link between domain data and domain knowledge and map the scope and scale of the field of informatics. Using acquisition, representation, processing, interpretation, and transformation, informatics transforms primitive data into useful knowledge.

As all types of information assets become online assets, malware (malicious software) has kept pace, developing into a huge source of cyber threats, either in the form of cyber attacks or cyber exploitation. At the CTO Roundtables organized by the ACM to define and articulate consensus best practice on newly emerging commercial technologies, experts were quoted as saying they have seen a huge explosion in the volume of new malware, and an evolution in terms of malware sophistication [30]. They have also seen a host of new data-mining techniques and new methods of self-protection in the cyber threat landscape. Cyber war in the fifth domain, cyberspace, has many battlegrounds (refer to Figure 1). Cyber defense and deterrence have at least three fronts: technology in the cyber world, operations in the physical world, and experience in the natural world applied to the three layers of the CPN ecosystem: Computing Utility (U), Interconnection Network (N) and Information Marketplace (M). Each of these nine battlegrounds (C, P, N) × (U, N, M) is interconnected to and interacts with eight others (refer to Figure 1).

In each of the nine subdomains (or battlegrounds) of cyber security, the goal is to be able to recognize patterns in a large set of data, to integrate information from multiple sensors and sources, and to extract useful knowledge from actionable insights (see for example [11, 19]). Various methods and techniques include agent-based computational modeling, forecasting algorithms that acknowledge uncertainty, and network theory that regulates systemic risk [4, 37, 117]. Mapping, evidence design, and visualization are also used [105, 106]. Data-mining, information fusion, and dependency-based region partitioning are used to detect intrusion in the distributed interconnection network [10, 20, 72]. Dawn Song, a MacArthur Foundation fellow, is helping cybercitizens police the Internet by analyzing data flow patterns of how software, hardware, and networks interact, and understand the flow of both "good" data and ill-intentioned hacks [28]. Her work contributes to the basis of two important platforms in cyber security: BitBlaze, which analyzes malicious software code, and WebBlaze, which focuses on the defense of Web-based applications and services against any attack or exploitation.

One of the important issues in cyber security is *trust*. Trust in various domains such as data, cloud computing, and e-trust is an important issue [82]. Trust in hardware and software is also crucial. A new hardware Trojan taxonomy is proposed to provide better understanding of existing and potential cyber threats [64]. On the software side, a study that compares the effectiveness of six commercial antivirus programs against malware was empirically conducted [100]. Another dimension of cyber security is *prevention*. Most research in cyber security has focused on defense. But practical cyber security measures require a balance between defense and deterrence. A network layer capability called *privacy-preserving forensic attribution* has been introduced to achieve such a balance [3]. Computer forensics, e-discovery, and evidence-based technology-assisted document review are described and proposed [48, 99].

Trust in hardware is a fundamental issue in cyber security. In an *ACM Ubiquity* interview in March 2002 [1, 70], the innovative computer scientist Ruby Lee talked about secure information processing using bit permutations, fair use in the digital age, and other important issues such as building security into the core software and hardware to provide another level of generic defense. In a preventive and proactive sense, Lee aims to build efficient, effective, and robust mechanisms for handling emerging cyber security and new multimedia directly into the computer architecture rather than heaping on more software or installing patches whenever there is a security problem [1, 21].

In order to provide a secure CPN cyberspace ecosystem, an evidence-based data-centric approach, as I mentioned earlier in this section, has to be taken. The data acquisition process includes data collection and data curation in the databases. Curated data can be studied or investigated through a scientific process, *informatics*, for representation, processing, and interpretation of information. The resulting information can be further analyzed using modeling, simulation, and visualization [55]. This pipeline has become a framework for the field of cyber security informatics. The processing step in the informatics stage involves mathematical, computational, and statistical methods and practices such as multiple comparisons, combining pattern classifiers, combining artificial neural nets, and combinatorial fusion analysis [8, 56, 69, 97]. All these emphasize the process of data analytics, information fusion, and knowledge discovery using the methods of computing, informatics, and analytics. Cyber security informatics has become an emerging scientific field.

Various machine learning and data-mining techniques can be applied to complex problems in cyber security [83, 119]. These include classification, association rules, clustering, decision trees, and random forest. Multiple classifier systems, multiple neural nets, and multiple scoring systems have recently been used to integrate data, combine features, and fuse decisions. They have been shown to have tremendous promise in solving complex problems for structured and unstructured data [56, 69, 97]. When combining multiple classifier systems, called *ensemble* in many cases, one aims to predict combination performance in terms of the performance of and diversity among individual systems. Since performance of individual systems and diversity among them may be correlated with each other, the prediction problem becomes extremely challenging for the researchers and practitioners. However, lower and upper bounds based on particular combination methods, performance evaluation, and diversity measures have been obtained to enable better situation analysis, decision making, and action taking [23, 24].

COLLABORATION AMONG GOVERNMENT, ACADEMIA AND INDUSTRY

In the combined cyber-physical-natural world, all the nine subdomains are interconnected and interrelated to each other (refer to Figure 1). Cyber security in the combined cyberspace has become a challenging issue in terms of technology, operation, and experience. Brenner [13] evaluated governmental actions (or inactions) to improve cyber security and raised a question: "Why isn't cyberspace more secure?" Despite three presidential directives to fix the problem [14, 31, 104], it is still easy to get away with cyber fraud, easy to steal corporate intellectual property, and easy to penetrate government networks. Other than market-or-liability-driven improvement, Brenner [13] offers eight steps the government could take to improve Internet security without creating any new bureaucracy or intrusive regulations. Two examples are using purchasing power to require higher security standards of vendors and amending the Privacy Act to make it clear that Internet Service Providers (ISPs) must disclose to one another and to their customers when a customer's computer has become part of a botnet. It is also recommended that cross-departmental (or cross-ministerial) governance with an interdepartmental organ of directive power is needed to muscle-entrenched and parochial bureaucracies.

As I stated before, the combined cyberspace is a complex, dynamic, and interconnected digital infrastructure that touches all facets of the information society, including economy, communication, public safety, computing utility, networks, and marketplaces. Maughan [78] echoed the need for a national cyber security research and development agenda to provide technical and policy solutions that enable four initial aspects of trustworthy cyberspace: security, reliability, privacy, and usability. In the Cybersecurity Research Roadmap [2], broad research and development agenda requirements are addressed to enable production of enabling technologies that will protect future information systems and networks. The top ten list of the eleven hard problem areas in cyber security provided by the document for research development agenda consists of the following: (1) software assurance, (2) metrics, (3) useable security, (4) identity management, (5) malware, (6) hidden threats, (7) hardware security, (8) data provenance, (9) trustworthy systems, and (10) cyber economics [78].

Brenner [13] and Groth and MacKie-Mason [51] also emphasize two possible key partnerships for the government to successfully secure the future cyberspace, including its utilities, networks, and marketplaces. The first partnership is with the educational systems in the academic world. The Taulbee Survey [103] has indicated that our current academic world is not producing enough cyberspace workers for the future, in particular with computer science (CS) and science, technology, engineering, and mathematics (STEM) degrees. At the same time, STEM and CS jobs are growing faster than the national average. The second partnership is with the private sectors. This public-private partnership is crucial when we address cyber security issues because most of the information and communications networks are owned and operated by the private sectors, both nationally and internationally.

Collaboration between academic and private industry is also important for cyber security and the study of cyber security informatics. Verel [108] reported on a forum in March 2011 hosted by Fordham University and IBM to address "Analytic Skills for 21st Century Leaders of the Smarter Planet." Leaders from academia, industry, and government gathered at the

Fordham IBM Forum on Analytics to discuss how cutting-edge computing, emerging technologies, and data analytics could dramatically change the cyberspace landscape. It also echoed a talent shortage that could significantly affect the well-being of the CPN ecosystem, including, among others, energy, education, environment, health, health care, security, and global competitiveness.

International cooperation is vital for all countries in cyberspace. One attempt in this direction was a gathering of a group of specialists and diplomats from fifteen countries (Belarus, Brazil, Britain, China, Estonia, France, Germany, India, Israel, Italy, Qatar, Russia, South Africa, South Korea, and the United States) that was chartered by the United Nations in 2005 to explore various problems in international cyber security. Little progress had been made until July 2010. Markoff [77] reported that the group of specialists from the fifteen countries agreed on a set of recommendations to the United Nations Secretary General for negotiation on an international computer security treaty. Veni Markovski, representing the international organizations ICANN, applauded the effort and the accomplishment of such cooperation of many countries in the field of cyber security.

EDUCATION AND SUSTAINABILITY

In order to secure cyberspace in the twenty-first century, public and private partnerships and collaboration between government and academia are crucial. Partnerships between academia and private industry are also important for innovative solutions and real-world applications. Overall, in a secure cyberspace ecosystem, collaboration among government, academia and industry has become an essential ingredient for future success. Training of a cyberspace workforce and education of general cybercitizens can provide defense against cyber attacks and cyber exploitation. For example, the New York Metro InfraGard chapter organized a real-world cyber defense exercise for network dependents and attackers to hone their skills in a closed exercise network in a real-time digital cyberspace environment [58]. Several formal degrees and certificate programs have been recently organized in different academic institutions. These include, among others, the Security Studies Program (SSP) at Georgetown University's School of

Foreign Service [16], the MS in Cybersecurity program at the Polytechnic Institute of New York University [90], and Fordham University's wide range of courses and projects in cyber security informatics offered by the Department of Computer and Information Science [36].

Following the spirit of the fourth paradigm we mentioned earlier, the subject of informatics has been developed widely in the academic world. Degrees in informatics, information science, or x-informatics have been offered at the undergraduate and graduate (master's and PhD) level [51]. As a result, the field of cyber security informatics has become an emerging new field of academic study and research. It will generate a variety of relevant programs and projects in the world. The subject of cyber security informatics, which studies, among other things, data collection, curation, computing, modeling, and visualization, has been applied to the nine subdomains of the combined cyberspace ecosystem (refer to Figure 1).

In order to sustain a secure cyberspace ecosystem, we mention four areas to be considered seriously: accessibility, privacy, policy, and process management. The cyberspace ecosystem can be sustainable only if each of the citizens in the natural world has equal and fair access to the computing utilities, interconnection networks, and information marketplace. In this regard, Web accessibility must ensure the development of Web resources that all people can use or have access to, regardless of their technical, physical, or cognitive limitations [84]. New technologies and guidelines including directives, legislation, and standards have been developed to increase accessibility, in particular for people with disabilities. The International Organization for Standardization's ISO 9241-171 defines *accessibility* as the "usability of a product, service, environment or facility by people with the widest range of capabilities" [88, 114].

Privacy in the ecosystem of cyberspace is also an important issue. As the Internet becomes ubiquitous (so people anywhere can access online public databases and up-to-date information), it also poses great risk to user privacy, since a malicious database owner may monitor user queries and infer the user's intention. Yekhanin [120] discussed Private Information Retrieval (PIR) schemes, which are cryptographic protocols to safeguard the privacy of user queries to public databases. More specifically, a special class of error-correcting codes, called *locally decodable codes* (LDCs), can be used.

These protocols ensure that every individual server (by observing only the query it receives) gets no other information about the identity of the item of the user's interest. Schwartz [95] connects privacy to identity management and elaborates on an identity ecosystem which is related to multiple-trust frameworks.

Policy is an emerging area that will play an important role in securing the cyberspace ecosystem. It is also crucial in the collaboration among government, academia, and industry, and in the relationship between and among geographically located nations and states. A study by the National Research Council [86] in April 2009 focused on discussing various legal dimensions of cyber attack policy in the United States. The report distinguished cyber attack from cyber exploitation. Cyber attack is seen as an aggressive action to alter, disrupt, degrade, deceive, or destroy adversary computer systems, transmitting networks, or information in or generated from these systems and networks. Cyber exploitations consist of quiet actions to obtain information from adversary IT systems, software, and services. Cyber attacks happen at a variety of ranges from remote to close-access attacks; examples include viruses, supply chain attacks, operational compromise, and denial-of-service attacks. Eavesdropping and installation of a Trojan horse are two examples of cyber exploitation. The report also suggests the establishment of a cybercrime convention that aims to enhance international cooperation in investigation and prosecution [86]. Although there are limits and constraints on what a cybercrime convention can do, such a convention can at least lay out the cybercrime boundaries and provide some legal paradigms for a secure cyberspace ecosystem.

The fourth area for a secure and sustainable cyberspace ecosystem is the process management capability of organizations and governments. An organization's infrastructure has to be brought up to a level that can face the challenges and opportunities of a combined cyber world. One example of the necessary reorganization in government is the National Institute of Information and Communications Technology (NICT) in Japan, an incorporated administration agency launched in April 2004 as a merged entity from the Communications Research Laboratory (CRL) and the Telecommunications Advancement Organization of Japan (TAO) [59]. In April 2006, NICT articulated a new five-year plan to integrate its content of existing research and development into three research domains: new generation network architecture technology, universal communication basic technology, and ICT for safety and security. The formation of the U.S. CyberCommand was another recent event in the reorganization process responding to the emerging needs of the CPN cyberspace ecosystems.

Conclusion

In this overview, I first gave a review of the emerging cyberspace in terms of the digital landscape, the technology capacity, the fifth domain, and the fourth paradigm. Then a cyberspace infrastructure was described in terms of ubiquitous networks, internetworking, Internet Protocol address and IPv6, pervasive systems, and social interaction networks. Due to the immersion of a combined CPN cyberspace ecosystem, I propose to build a secure cyberspace ecosystem using cyber security informatics and utilizing the collaboration of all stakeholders in government, academia, and industry. In order for the secure cyberspace ecosystem to function efficiently and effectively in the long term, education and training of the stakeholders are needed. Four ingredients of the secure cyberspace—accessibility, privacy, policy, and process management—are recommended to sustain the secure cyberspace ecosystem.

In essence, a combined CPN cyberspace is the result of fusion from three diverse existing worlds: the cyber world, the physical world, and the natural world. With the immersion of these three worlds and the interaction between the three layers of computing utility, interconnection network, and information marketplace, the cyberspace ecosystem consists of the nine worlds functioning at three levels (refer to Figure 1). The cyberspace ecosystem is really a complex system interconnecting and functioning as an ensemble of nine subdomains. In this overview, I did not intend to give a complete account of the dynamic and evolving cyberspace. Nor did I intend to give any set of complete solutions to secure the emerging ecosystem. It was aimed to provide a holistic view of the emerging complex cyberspace, to offer an integrated approach using the combination of computing, informatics, and analytics to secure the complex ecosystem, and to suggest ways to sustain the secure cyberspace ecosystem.

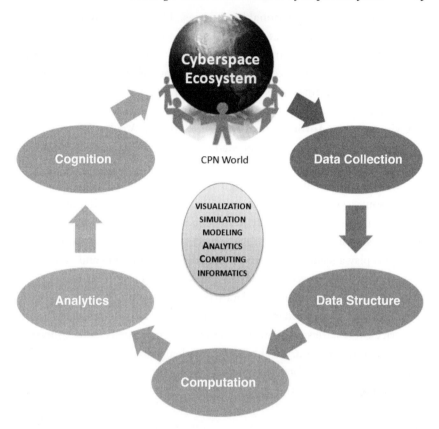

Figure 2. Cybersecurity Informatics Framework (CIF) for the CPN cyberspace ecosystem.

I hope this overview will serve as a catalyst for all stakeholders of the cyberspace to conceive novel ideas and construct innovative solutions to address micro- and macro-challenges and opportunities for a secure cyberspace ecosystem. I also recommend the establishment of an International Consortium of Cybersecurity Professionals and the Cybersecurity Informatics Framework:

International Consortium of Cyber Security Professionals (ICCP):
An international consortium is needed to coordinate cyber security professionals from government, academia, and industry across

national and international boundaries. These professionals are researchers, teachers, and practitioners working on technology, operation, and domain applications. They would share their practical experiences, strategize their functional operations, and utilize their innovative technologies to combat cyber attack and cyber exploitation.

Cybersecurity Informatics Framework (CIF): A cyber security informatics framework is needed to survey the current ecosystem of a secure cyberspace such as databases, tools, groups, materials, and resources. The goal is to design and develop a cyber security resource system (CRS) and establish an efficient and effective search strategy for cyber security in the cyberspace ecosystem to locate, access, and retrieve digital cyber security-related resources. It will also play a leading role as a portal site for any matters and affairs related to cyber security. Using the concept of the informatics pipeline—data collection, data structure, computation, analytics, and cognition, similar to the one mentioned in [55]—a Cybersecurity Informatics Framework (CIF) for the cyberspace ecosystem is illustrated in Figure 2. This framework also includes computing, analytics, and informatics methods together with modeling, simulation, and visualization systems [68].

REFERENCES

[1] "A conversation with Ruby Lee," ACM Ubiquity interview, *Ubiquity*, March (2002).

[2] A Roadmap for Cybersecurity Research (2009, November). Department of Homeland Security, Science and Technology Directorate. [Online]. Available: http://www.cyber.st.dhs.gov/resources/

[3] M. Afanasyev, T. Kohno, J. Ma, N. Murphy, S. Savage, A. C. Snoeren, and G. M. Voelker, "Privacy-preserving network forensics," *Communications of the ACM*, Vol. 54, No. 5, May (2011), pp. 78–87.

[4] M. Allen, "Embracing an uncertain future," *Nature*, Vol. 466, July (2010), p. 31.

[5] G. Anthes, "Security in the Cloud," *Communications of the ACM*, Vol. 53, No. 11 (2011), pp. 16–18.

[6] M. Armbrust, A. Fox, R. Griffith, A. D. Joseph, R. Katz, A. Konwinski, G. Lee, D. Patterson, A. Rabkin, I. Stoica, and M. Zaharia, "A view of

cloud computing," *Communications of the ACM*, Vol. 53, No. 4 (2010), pp. 50–58.

[7] M. Baker, "From promising to practical: tools to study networks of neurons," *Nature Methods*, Vol. 7, No. 11 (2010), pp. 877–883.

[8] N. Balakrishnan, N. Kannan, and H. N. Nagaraja (Eds), *Advances in Ranking and Selection, Multiple Comparisons, and Reliability*, Birkhauser (2005), 412 pages.

[9] A.-L. Barabasi, *Linked*, Plume Book (2003).

[10] T. Bass, "Intrusion detection systems and multisensor data fusion," *Communications of the ACM*, Vol. 43, No. 4, April (2000), pp. 99–105.

[11] "Better business outcomes with business analytics," IBM Software Group, IBM Corporation, October (2010).

[12] I. Bremmer, "Democracy in cyberspace," *Foreign Affairs*, Vol. 89, No. 6 (2010), pp. 86–92.

[13] J. F. Brenner, "Why isn't cyberspace more secure?" *Communications of the ACM*, Vol. 53, No. 11, November (2010), pp. 33–35.

[14] President G.H.W. Bush's "National Security Directives," July 5 (1996).

[15] M. Castells, "End of Millennium," *The Information Age: Economy, Society and Culture*, Vol. III, Wiley-Blackwell, Malden, MA (2000).

[16] Center for Peace and Security Studies, Georgetown University. [Online]. Available: http://ssp.georgetown.edu

[17] V. Cerf and R. E. Kahn, "A protocol for packet network intercommunications," *IEEE Transactions on Communications*, Vol. 22, No. 5 (1974), pp. 637–648.

[18] V. Cerf, Y. Dalal, and C. Sunshine, 1RFC675. "Specification Internet Transmission Control Program" (1974).

[19] "Challenges and Opportunities," *Science*, Vol. 331, No. 6018, February 11 (2011), pp. 692–693.

[20] H. Chen, W. Chung, J. J. Xu, G. Wang, Y. Qin, and M. Chau, "Crime data mining: a general framework and some examples," *Computer*, IEEE Computer Society, Vol. 37, No. 4, April (2004), pp. 50–56.

[21] A. Chen, "Professor calls for new approaches in improving cyber security," *INSIDE Fordham*, August (2010).

[22] N. Christakis and J. Fowler, *Connected: The surprising power of our social networks and how they shape our lives*, New York: Little, Brown and Co. (2009).

[23] Y.-S. Chung, D. F. Hsu and C.-Y. Tang, "On the diversity-performance relationship for majority voting in classifier ensembles, in Multiple Classifier Systems," *LNCS 4472* (edited by M. Haindl, J. Kittler, and F. Roli), Springer-Verlag (2007), pp. 407–420.

[24] Y.-S. Chung, D. F. Hsu, C.-Y. Liu and C.-Y. Tang, "Performance evaluation of classifier ensembles in terms of diversity and performance of individual systems," *International Journal of Pervasive Computing and Communications*, Vol. 6, No. 4 (2010), pp. 373–403.

[25] W. K. Clark and P. L. Levin, "Securing the information highway; How to enhance the United States' electronic defenses," *Foreign Affairs*, Vol. 88, No. 6 (2009), pp. 1–10.

[26] R. A. Clarke, *Cyber War: The Next Threat to National Security and What to Do About It*, HarperCollins, New York, NY (2010), 290 pages.

[27] Cooperative Association for Internet Data Analysis (CAIDA).

[28] M. V. Copeland, "A genius approach to web security," *Fortune*, March 21 (2011), p. 52.

[29] T. M. Cover and J. A. Thomas, *Elements of Information Theory*, Wiley-Interscience, Hoboken, NJ (2006).

[30] M. Creeger, "CTO Roundtable: Malware defense," *Communications of the ACM*, Vol. 53, No. 4 (2010), pp. 43–49.

[31] Cyberspace Policy Review: Assuring a Trusted and Resilient Information and Communication Infrastructure (2009, May). [Online]. Available: http://www.whitehouse.gov/assets/documents/Cyberspace_Policy_Review _final.pdf

[32] "Cyberwar," *The Economist*, July 3–9 (2010), pp. 1–12.

[33] J. D. Day and H. Zimmermann, "The OSI Reference Model," *Proc. of the IEEE*, Vol. 71 (1983), pp. 39–55.

[34] "Dealing with data-challenges and opportunities," *Science*, Vol. 331 (2011), pp. 692–693.

[35] R. DeCastro and J. W. Grossman, "Famous trails to Paul Erdös," *Mathematical Intelligence*, Vol. 21 (1999), pp. 51–63.

[36] Department of Computer and Information Science, Fordham University. [Online]. Available: http://www.cis.fordham.edu/graduate.html

[37] D. Diermeier, "Arguing for computational power," *Science*, Vol. 318, No. 5852, November 9 (2007), pp. 918–919.

[38] A. Doan, R. Ramakrishnan, and A. Y. Halevy, "Crowdsourcing systems on the World-Wide Web," *Communications of the ACM*, Vol. 54, No. 4, April (2011), pp. 86–96.

[39] S. Dutta, "What's your personal social media strategy?" *Harvard Business Review*, Nov. (2010), pp. 127–130.

[40] C. Fass, M. Ginelli, and B. Turtle, *Six degrees of Kevin Bacon*, Plume Book (1996).

[41] J. Fine, "Can harvesting data make better wine?" Food and Wine.com (2011, June), pp. 58–62. [Online]. Available: http://www.foodandwine.com/articles /wine-technology-can-harvesting-data-make-better-wine

[42] J. F. Gantz et al. (2008). "The diverse and exploding digital universe: An updated forecast of worldwide information growth through 2011," International Data Corporation White Paper sponsored by EMC. [Online]. Available: http://www.emc.com/collateral/analyst-reports/diverse-exploding -digital-universe.pdf

[43] M. Gladwell and C. Shirky, "From innovation to revolution: Do social media make protests possible?" *Foreign Affairs*, March/April (2011), pp. 153–154.

[44] A. Goldfarb and C. E. Tucker, "Online Advertising, Behavioral Targeting, and Privacy," *Communications of the ACM*, Vol. 54, No. 5, May (2011), pp. 25–27.

[45] G. Goth, "Turning data into knowledge," *Communications of the ACM*, Vol. 53, No. 11, Nov. (2010), pp. 43–49.

[46] S. Greengard, "Social games, virtual goods," *Communications of the ACM*, Vol. 54, No. 4, April (2011), pp. 19–21.

[47] D. A. Grier, "Not for all markets," *Computer*, IEEE Computer Society, Vol. 44, No. 5, May (2011), pp. 6–8.

[48] M. R. Grossman and G. V. Cormack (2011). "Technology-assisted review in E-Discovery can be more effective and more efficient than exhaustive manual review," *Richmond. Journal of Law and Tech*, Vol. 1, No. 48 (2011). [Online]. Available: http://jolt.richmond.edu/v17i3/article11.pdf

[49] J. W. Grossman, The Erdös Number Project. [Online]. Available: http://www.oakland.edu/enp/

[50] L. Grossman, "The time traveler," *Time*, September 13 (2010), pp. 62–63.

[51] D. P. Groth and J. K. MacKie-Mason, "Why an informatics degree?" *Communications of the ACM*, Vol. 53, No. 2, Feb. (2010), pp. 26–28.

[52] J. Guare, *Six Degrees of Separation*, Random House, New York (1990).

[53] T. Hey et al. (edited), "Jim Gray on e-Science: A Transformed Scientific Method, in *The Fourth Paradigm*," Microsoft Research (2009), pp. 19–31.

[54] M. Hilbert and P. Lopez, "The world's technological capacity to store, communicate and compute information," *Science*, Vol. 332, April 1 (2011), pp. 60–65.

[55] D. F. Hsu, B. S. Kristal, C. Schweikert, "Rank-score characteristic (RSC) function and cognitive diversity," *Brain Informatics*, LNAI 6334, Springer (2010), pp. 42–54.

[56] D. F. Hsu, Y. S. Chung, and B. S. Kristal, "Combinatorial fusion analysis: methods and practices of combining multiple scoring systems," in H. H. Hsu (Ed.), *Advanced Data Mining Technologies in Bioinformatics*, Idea Group, Hershey, PA (2006), pp. 32–62.

[57] D. F. Hsu, "On container width and length in graphs, groups, and networks," *IEICE Trans. Fund. Elect., Commun. Comput. Sci.* E77-A (4) (1994), pp. 668–680.

[58] InfraGard Cyber Defense Initiative 2009, New York Metro InfraGard Chapter, July 21–22 (2009).

[59] National Institute of Information and Communications Technology (NICT). [Online]. Available: http://www.nict.go.jp/about/message-e.html

[60] A. Ivanov, "Using prediction markets to harness collective wisdom for forecasting," *Journal of Business Forecasting*, Vol. 28, No. 3 (2009), pp. 9–14.

[61] J. Jacobs, *The Death and Life of American Cities*, Random House, New York, (1961).

[62] V. Jacobson and M. Karels, "Congestion Avoidance and Control," *Proceedings of SIGCOMM '88*, Sept. (1988), pp. 314–329.

[63] D. B. Johnson, "Scalable support for transparent mobile host internetworking," *Wireless Networks*, Vol. 1 (1995), pp. 311–321.

[64] R. Karri, J. Rajendran, K. Rosenfeld, and M. Tehranipoor, "Trustworthy hardware: Identifying and classifying hardware trojans," *IEEE Computer Magazine*, Vol. 43, No. 10 (2010), pp. 39–46.

[65] P. Key, L. Massoulié, and D. Towsley, "Path selections and multipath congestion control," *Communications of the ACM*, Vol. 54, No. 1 (2011), pp. 109–116.

[66] J. Kleinberg, "Authoritative sources in a hyperlinked environment," *Journal of the ACM*, Vol. 46 (1999), pp. 604–632.

[67] V. Kostakos and P. Kostakos, "Inferring social networks from physical interactions: a feasibility study," *International Journal of Pervasive Computing and Communications*, Vol. 6, No. 4 (2010), pp. 423–431.

[68] M. E. Kuhl, J. Kistner, K. Costantini, and M. Sudit, "Cyber attack modeling and simulation for network security analysis," *Proceedings of IEEE 2007 Winter Simulation Conference* (edited by S.G. Henderson et al.) (2007), pp. 1180–1188.

[69] L. I. Kuncheva, *Combining Pattern Classifiers: Methods and Algorithms*, John Wiley & Sons, Hoboken, NJ (2004).

[70] R. B. Lee, Z. Shi, and X. Yang, "Efficient permutations for fast software cryptography," *IEEE Micro*, Vol. 21, No. 6, December (2001), pp. 56–69.

[71] M. Lesk. (1977). "How much information is there in the world?" [Online]. Available: available at http://www.lesk.com/mlesk/ksg97/ksg.html

[72] J. Li, K. Sollins, D.-Y. Lim, "Dependency-based distributed intrusion detection system," *Proceedings of the DETER Community Workshop on Cyber Security Experimentation and Test (2007)*.

[73] T. A. Limoncelli, "Successful strategies for IPv6 rollouts: Really," *Communications of the ACM*, Vol. 54, No. 4, April (2011), pp. 44–48.

[74] Y.-Y. Liu, J.-J. Slotine, and A.-L. Barabasi, "Controllability of complex networks," *Nature*, Vol. 473, May (2011), pp. 167–173.

[75] S. Lloyd, *Phys. Rev. Lett. 88: 237901 (2002)*.

[76] J. L. Gómez-Barroso and C. Feijóo, "Asymmetries and shortages of the network neutrality principle," *Communications of the ACM*, Vol. 54, No. 4, April (2011), pp. 36–37.

[77] J. Markoff, "Step taken to end impasse on cybersecurity talks," *The New York Times*, July 16 (2010).

[78] D. Maughan, "The need for a national cybersecurity R&D agenda," *Communications of the ACM*, Vol. 53, No. 2, Feb. (2010), pp. 29–31.

[79] D. J. McFarland and J. R. Wolpaw, "Brain-computer interfaces for communication and control," *Communications of the ACM*, Vol. 54, No. 5, May (2011), pp. 60–66.

[80] T. Mikkonen and A. Taivalsaari, "Reports of the Web's death are greatly exaggerated," *Computer*, IEEE Computer Society, Vol. 44, No. 5, May (2011), pp. 30–36.

[81] S. Milgram, "The small world problem," *Physiology Today*, Vol. 2 (1967), pp. 60–67.

[82] K. W. Miller, J. Voas, and P. Laplante, "In trust we trust," *Computer*, IEEE Computer Society, Vol. 43, No. 10, Oct. (2010), pp. 85–87.

[83] T. Mitchell, *Machine Learning*, McGraw Hill, 1997.

[84] L. Moreno, P. Martinez, B. Ruiz, and A. Iglesias, "Toward an equal opportunity web: Applications, standards, and tools that increase accessibility," *Computer*, IEEE Computer Society, Vol. 44, No. 5, May (2011), pp. 18–26.

[85] P. Norvig, "Search, in 2020 Visions," *Nature*, Vol. 463, January 7 (2010), p. 26.

[86] W. A. Owens, K. W. Dam, and H. S. Lin (Eds.), "Technology, Policy, Law, and Ethics Regarding U.S. Acquisition and Use of Cyberattack Capabilities," Committee on Offensive Information Warfare, National Research Council, 2009.

[87] L. D. Paulon, "Dual core processor to improve smartphone performance," *Computer*, IEEE Computer Society, Vol. 44, No. 5, May (2011), p. 16.

[88] L. D. Paulson, "W3C adopts web-accessibility specifications," *Computer*, IEEE Computer Society, Feb. (2009), pp. 23–26.

[89] C. Pawela and B. Biswal, "Brain Connectivity: A New Journal Emerges," *Brain Connectivity*, Vol. 1, No. 1 (2011), pp. 1–2.

[90] Polytechnic Institute of New York University, http://poly.edu/graduate

[91] I. D. S. Pool, "Tracking the Flow of Information," *Science 221* (1983), pp. 609–613.

[92] N. Ramakrishnan, D. Hanauer, and B. Keller, "Mining electronic health records," *Computer*, IEEE Computer Society, Vol. 43, No. 10, Oct. (2010), pp. 77–81.

[93] E. Schadt, "Molecular networks as sensors and drivers of common human disease," *Nature*, Vol. 461, September 10 (2009), pp. 218–223.

[94] E. Schmidt and J. Cohen, "The digital disruption: Connectivity and the diffusion of power," *Foreign Affairs*, Vol. 89, No. 6 (2010), pp. 75–85.

[95] A. Schwartz, "Identity management and privacy: a rare opportunity to get it right," *Communications of the ACM*, Vol. 54, No. 6 (2011), pp. 22–24.

[96] C. E. Shannon, "A mathematical theory of communication," *Bell. Syst. Tech. J.* 27 (1948), pp. 319–423 and 623–656.

[97] A. Sharkey (Ed.) *Combining Artificial Neural Nets*, Springer-Verlag, Berlin, (1999).

[98] C. Shirky, "The political power of social media," *Foreign Affairs*, Vol. 90, No. 1, January-February (2011), pp. 28–41.

[99] E. M. Stroz and D. F. Hsu, "Improving knowledge discovery for the justice system through informatics," *IEEE ITProfessional*, (2012), pp. 47–52.

[100] O. Sukwong, H. S. Kim and J. C. Hoe, "Commercial antivirus software effectiveness: an empirical study," *IEEE Computer*, March (2011), pp. 63–70.

[101] A. S. Tanenbaum, *Computer Networks*, Fifth Edition, Prentice-Hall PTR, (2010).

[102] J. C. Tang, M. Cebrian, N. A. Giacobe, H.-W. Kim, T. Kim, and D. Wickert, "Reflecting on the DARPA red balloon challenge," *Communications of the ACM*, Vol. 54, No. 4, April (2011), pp. 78–85.

[103] "Taulbee Survey 2006–2007," *Computing Research News*, Vol. 20, No. 3, Computing Research Association, May, 2008.

[104] "The National Strategy to Secure Cyberspace 2003," U.S. Department of Homeland Security.

[105] E. R. Tufte, *Envisioning Information*, Cheshire, CT: Graphics Press, May (1990).

[106] E. R. Tufte, *The Visual Display of Quantitative Information*, Second Edition, Cheshire, CT: Graphics Press, January (2001).

[107] A. M. Turing, "On computable numbers, with an application to the entscheidungs problem," *Proc. London Math. Soc.*, Vol. s2-42, No. 1 (1937), pp. 230–265.

[108] P. Verel, "Analytics conference shows possibilities of collaboration," *INSIDE Fordham*, April 26 (2011).

[109] "War in the fifth domain," *The Economist*, July 3–9 (2010), pp. 25–28.

[110] S. Wasserman and K. Faust, *Social Network Analysis: Methods and Applications*, Cambridge University Press (1994), pp. 598–602.

[111] D. J. Watts and S. H. Strogatz, "Collective dynamics of 'small-world' networks," *Nature*, Vol. 393 (1998), pp. 440–442.

[112] D. J. Watts, *Six Degrees: The Science of a Connected Age*, W.W. Norton & Company, New York (2004).

[113] "Weaving a New Web," *Discover*, March (2011), pp. 54–60.

[114] Web Accessibility Initiative (WAI). [Online]. Available: http://www.w3.org/WAI

[115] M. Welsh, "Sensor networks for the sciences," *Communications of the ACM*, Vol. 53, No. 11 (2011), pp. 36–39.

[116] D. Wischik, "Multipath: a new control architecture for the internet," *Communications of the ACM*, Vol. 54, No. 1 (2011), p. 108.

[117] C. Wright, "Six degrees of rumination," *CFA Magazine*, Vol. 20, No. 4, July/August (2009), pp. 42–45.

[118] A. Wright, "Web science meets network science," *Communications of the ACM*, Vol. 54, No. 5, May (2011), p. 23.

[119] X. Wu and V. Kumar (Eds), *The Top Ten Algorithms in Data Mining*, Chapman & Hall/CRC, 2009.

[120] S. Yekhanin, "Private information retrieval," *Communications of the ACM*, Vol. 53, No. 4, April (2010), pp. 68–73.

Part I: Technology

Improving Cyber Security

Ruby B. Lee

Cyber security is essential given our growing dependence on cyberspace for all aspects of modern societies. However, today, attackers have the upper hand. In this chapter, I discuss the security properties needed, and some key strategies that may have the potential to level the playing field between attackers and defenders. These research strategies were developed at the National Cyber Leap Year summit, with experts from industry, academia, and government working collaboratively. These broad research thrusts can be interpreted at different levels of the system, and in different application domains. Because a promising direction explored at the summit is the use of hardware architecture to enhance security, I provide a hardware-enhanced interpretation of the proposed research thrusts, with the goal of illustrating how new security features can be built into future commodity computers to improve system security. The goal is to be able to ensure essential security features for critical tasks, even in the presence of malware and software vulnerabilities in the system, and users who are not security-savvy. Feasibility examples are given to show how new hardware security features can help improve software security and also how hardware itself can be designed to be more trustworthy. These examples illustrate that by rethinking the fundamental design of computers with security as one of the key requirements, we can design future *secure, trustworthy computers* without necessarily sacrificing performance and other goals.

Cyber Security Today

In its early days, the Internet was used as a means of enhancing research among collaborating scientists. Internet protocols were designed to enable seamless *inter-network* communications between sender and receiver across heterogeneous networks, resulting in the name "Internet." Researchers worked hard to define the basic Internet protocols[1] and make them work, hiding the physical complexity of the different physical network technologies being used while allowing full flexibility to implement arbitrary applications across heterogeneous networks. The success of this design can be seen today in the various applications built on top of the Internet, such as the World Wide Web, search, web mail, e-commerce, e-banking, and social network applications. Today, our social lives, our economic competitiveness, our national security, and in fact all aspects of our lives depend on the correct functioning and ubiquitous availability of the Internet and wireless networks. This dependence on the Internet and on cyberspace transactions is increasing at the same time cyber attacks are escalating.

With society's growing dependence on cyberspace, cyber security becomes a critical issue. The technologies that make up cyberspace were not designed with security in mind. Internet protocols were designed for friendly parties to communicate and collaborate with each other—they were not designed with malicious adversaries in mind. Similarly, computer technology, both hardware and software, was not designed with attackers in mind. Hence, it should not be surprising that the basic network, soft-

---

1. For example, supporting the TCP/IP (Transmission Control Protocol/Internet Protocol) protocols enables the fundamental interoperability across different types of networks.

ware, and hardware technologies underlying cyberspace are full of security vulnerabilities that can be exploited by malicious parties. In addition, attackers only need to find one path into the system to infiltrate it, whereas defenders have to defend on all fronts. Furthermore, since most computer systems use the same systems software and hardware, the attacker will also have found a way to infiltrate a large majority of today's computers.

AN EXAMPLE OF AN INTERNET-SCALE ATTACK

An example of an Internet-scale attack is a *distributed denial-of-service* (DDoS) attack. In a DDoS attack, an army of zombies (also called a *botnet*) is harnessed to attack a *primary victim*, which could be a website, computer, or network (see Figure 1a). The zombies are actually *secondary victims*, since they are innocent computers whose owners are unaware that their machines have been infected with malware that can turn their machines into "zombies" (also called "bots" or "agents") that can be used in a DDoS attack on a primary victim.

Such stealthy infiltration of computers and installation of zombie programs can be done months before an actual DDoS attack is launched. They are possible due to security vulnerabilities, most often in the software, that can be exploited by an attacker to infiltrate a computer and silently install a zombie program without being detected. They even allow for updating such bot code in the zombie computer and sending the computer's configuration updates to the adversary. During a DDoS attack, the zombie programs are invoked to send innocuous requests to the primary victim website. Such requests are essentially indistinguishable from legitimate requests to this website. However, a flood of zombie requests from a large number of computers to a victim website can overwhelm it and its surrounding network, making it unable to provide service for legitimate requests (see Figure 1b). Hence, this is a denial-of-service (DoS) attack on the availability of services. It is called a *distributed* denial-of-service (DDoS) attack because the flood of requests comes from frontline zombie computers, which can be distributed all over the world, making it hard to drop traffic from just a few sources.

A DDoS attack is disturbing because it is like using innocent citizens without their consent to form an ad-hoc army to cause damage to one's

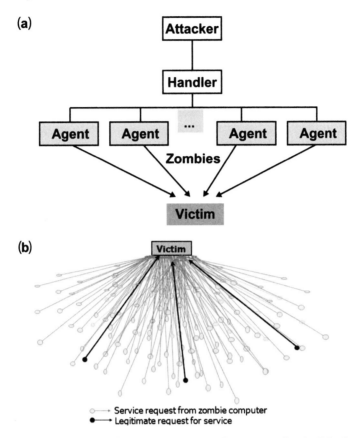

Figure 1. Distributed denial-of-service (DDoS) attack: (a) Example of a DDoS attack network; (b) DDoS attack flooding a primary victim site.

own country or other countries. Tracing the true attacker is difficult, since he will typically hide behind many levels of indirection, using *handlers* (which are themselves infected computers) to infiltrate other computers and install handler or zombie code on them. The zombie code and the actual requests made during a DDoS attack often use very little storage, computing, and networking resources, so they essentially run "under the radar" and are invisible to the user of the zombie computer. With software and networking enabled in trillions of devices and embedded systems, rather than just millions of computers, DDoS attacks can become orders of magnitude more potent in flooding network bandwidth and computer resources.

There are many security properties required or desired in computer and communications systems. Three such properties have been called *cornerstone* security properties: confidentiality, integrity and availability; their acronym, CIA, is the same as a famous U.S. agency and hence is easy to remember. *Confidentiality* is the prevention of the disclosure of secret or sensitive information to unauthorized users or entities. *Integrity* is the prevention of unauthorized modification of protected information without detection. *Availability* is the provision of services and systems to legitimate users when requested or needed. This is similar to providing reliability (or fault-tolerant systems), except that availability, in the security sense, is harder because it has to also consider intelligent attackers with malevolent intent whose behavior cannot be characterized by a probability distribution—unlike device failures and faults.

In addition to the fundamental CIA triad, there are many other important aspects of security, some of which I define briefly here. *Access control*, comprising both *authentication* and *authorization*, is essential to ensure that only authenticated users who are authorized to access certain information can in fact access that material, while denying access to unauthorized parties. *Attribution* support should ideally be built into secure systems in order to find the real attackers when a security breach has occurred. *Accountability* support holds vendors, owners, users, and systems responsible for vulnerabilities that enable successful attacks. *Non-repudiation* is desirable to ensure that a user cannot deny that he has made a certain request or performed a certain action. *Attestation* is the ability of a system to provide some non-forgeable trust evidence of its hardware and the software it is currently running. *Anonymity* is the ability to perform certain actions without being identified or tracked. *Privacy* is the right to determine how one's personal information is disseminated, or redistributed by an authorized recipient. Confidentiality and privacy are different in the sense that confidentiality is the obligation to protect secret information, while privacy is the right to protect one's personal information. However, it is possible that they may share similar defense mechanisms.

I define a *secure computer system* as one that provides at least the three cornerstone security properties of confidentiality, integrity and availability.

A *trustworthy computer* is one that is *designed* to be dependable and to provide such security properties, to do what it is supposed to do and nothing that may harm itself or others. In contrast, what has conventionally been proposed is a *trusted computer*—one that is depended on to enforce security policies, but if it is infiltrated, then all bets are off for enforcing security policies. More precisely, this has been described as a computer with a *trusted computing base* (TCB), which if violated means that security policies may not be enforced. Unfortunately, no COTS (commodity off the shelf) computer system today can achieve a dependable trusted computing base. In fact, commodity computers just have not been designed to be trustworthy. Security has typically been bolted on, after the fact, resulting in degradation of performance, cost, time-to-market, and ease-of-use. Furthermore, security issues are not even in the core computer architecture, hardware or software curriculum in colleges in the United States today, whereas they were more routinely taught thirty to forty years ago. This has created a crisis in educated manpower in the research, design and development of secure and trustworthy computer systems that are so sorely needed for cyber security. Hence, there is an urgent need to remedy this and to consider both the research and the educational needs for improving cyber security.

Strategies for Improving Cyber Security

Today, cyber attackers have a highly asymmetric advantage over defenders. How can we level the playing field so that attackers do not have such an upper hand? How can we increase the work factor for mounting successful cyber attacks by orders of magnitude?

Among many public and private sector initiatives, there was a government-sponsored initiative called the National Cyber Leap Year (NCLY) initiative, whose goal was to find game-changing strategies for improving cyber security, leaping ahead of current incremental or reactive strategies. Multiple rounds of calls for proposals for research directions were issued, and hundreds of responses came from both companies and universities. From these, a few major research directions were honed and discussed in an NCLY summit held in August 2009 to which security experts from industry and academia were invited. The summit was hosted by NITRD, representing thirteen government agencies, with many security experts from these government agencies present. It promoted constructive discussion of five research directions that were considered to have game-changing potential: hardware-enabled trust, digital provenance, nature-inspired cyber health, moving target defense, and cyber economics. The findings were summarized in the NCLY Summit Co-Chairs' Report [1]. While these approaches were considered to be promising directions where cyber security research is required, this does not in any way imply that other approaches are not promising.

The summit brought together experts from industry, academia, and government, with very different backgrounds and viewpoints, to rally under the common goal of improving cyber security. As an invited co-chair, I found the collaboration promoted by the summit quite refreshing, as did other co-chairs and participants. Indeed, such collaboration is needed to make real inroads in improving cyber security. More importantly, it is gratifying to note that this summit has had significant impact on government funding for cyber security research. Within about eighteen months, many government agencies, including NSF, DARPA, DHS, and DOD, to name a few, put out calls for proposals for research funding in these and related cyber security research areas.

Some of the promising research directions discussed at the summit can be combined under a broader umbrella of what might be termed "proactive design strategies" to:

enable tailored trustworthy spaces within the generally untrusted
    cyberspace;
thwart attackers with moving target defenses; and
reward responsible behavior with economic and other incentives.

These three major thrusts were presented at the NCLY kickoff meeting in May 2010 [2]. Below, I will first discuss these three strategies, in general, for improving cyber security. Then, to illustrate how these research strategies can be applied not only to the more visible layers of software and networking, but also to fundamental hardware design, I will give some feasibility examples on how to build future trustworthy computers that can help improve cyber security with hardware-enhanced trust.

ENABLING TAILORED TRUSTWORTHY SPACES

The goal of this research direction is to create technology that can provide trustworthy spaces on demand, tailored to the needs of the usage scenario.

This is in contrast to the conventional strategy where a *sandbox* is set up for executing untrusted applications. A sandbox is a constrained execution environment where the untrusted application is unable to use all the resources normally available on the computer system. While this may prevent it from accessing or modifying critical system resources, it also means that many existing applications will not run as they did before, leading to user dissatisfaction with such "security-upgraded" systems.

The fundamental difference of the proposed strategy is that untrusted applications run as before, with the same access to the system resources that they had before. However, trusted applications that are security-critical should be enabled to run in newly created *trustworthy spaces*, tailored to their needs. This recognizes that there is no one-size-fits-all trustworthy space. Instead, different trustworthy spaces should be created based on the security requirements of the particular application, or of the secure, sensitive, or proprietary data that are to be accessed. For example, the confidentiality levels required are different for protecting YouTube videos, movie rentals, online medical records, bank accounts and nuclear weapons. Similarly, confidentiality, integrity, availability, and privacy requirements differ from one application to another, from one environment to another, and from one use-case to another. Hence, the goal is to provide a secure execution environment for trusted applications without constraining untrusted applications, which can then run as before and also use the full functionality provided by commodity systems.

While this can be achieved in many ways, one approach is to set up a *secure execution compartment*, tailored to the security needs of the particular application, user, environment, and context. The security-critical part of an application then runs only in this secure execution compartment, where it is protected from other applications and also from system software such as the operating system or hypervisor, which might want to snoop on its confidential data or code. In addition to secure setup and secure execution, it is also important to provide secure termination of such secure execution compartments, to prevent an a posteriori leak of confidential information or contamination that may affect its future use.

An example of how small changes to the hardware and software architecture can enable this is discussed later, in the Bastion Security Architecture section.

In another approach, it would be great if we could enable *self-protecting data*—that is, data that can protect themselves from security breaches. This can be interpreted as data that have an attached security policy that cannot be violated no matter what application uses the data. Hence, the application does not have to be certified as trusted before it can be allowed to access protected data. Such data would have confidentiality and integrity policies that are either predefined, or can vary dynamically based on the context, the time, the location, the application, the system, and the user accessing the data. New hardware may be needed to enforce such policies, but for older computers without this new hardware, the self-protecting data might be accessible only in encrypted form, for example, with no access to the decryption key.

Data provenance, where the chronological history of the ownership of the data can be reliably obtained, is also highly desirable, both to establish data authenticity and to track adversaries when security has been breached.

It is also desirable for the system to be able to give some *trust evidence* to users that the protections they requested for their trustworthy spaces have indeed been set up and are being enforced. This may require the system to be able to perform *tailored attestations* [3] to give unforgeable assurances to the requester of the identity, type, or properties of the hardware and the software that is running on the system.

These are just a few examples of some promising research directions that may help enable the dynamic establishment of tailored trustworthy spaces within the untrusted cyberspace—without crimping the flexibility and functionality of existing applications or future applications that do not need such security.

It is not easy to set up such tailored trustworthy spaces, on demand, over the public Internet in insecure cyberspace. But if it can be done, it can enable the resilient execution of security-critical tasks, even in the presence of active attacks or malware previously introduced into the computer system.

THWARTING ATTACKERS WITH MOVING TARGET STRATEGIES

The goal of this research direction is to significantly increase the attacker's work factor for a successful attack and to reduce the number of machines that succumb to the same attack path.

One problem with today's homogeneous computing environments is that once an attacker finds a penetration path into a system, he can penetrate a huge number of similar systems. The majority of today's desktop and notebook computers use the Microsoft Windows operating system and Intel microprocessors, and the same web browsers, e-mail programs, and database software. Hence, many computers will have the same security vulnerabilities in either the software or the hardware. The idea of a moving target defense strategy is to have each system look different to an attacker, and even have a single system look different over time. This means that the attacker must learn to penetrate each system separately, and relearn how to penetrate a system he has already found a path into before—thus significantly increasing the attacker's work factor for a successful attack. A major challenge is to ensure that the system is as easy for legitimate users to use as it was before, while making it orders of magnitude harder for attackers. This moving target defense strategy also changes the game from today's reactive defense posture triggered by new attacks, to a preemptive posture where systems are specifically designed to look like moving targets to potential attackers.

A promising approach in moving target defense strategies is to use randomization in system design. This could potentially improve both the security and the performance of the system, rather than trading off one for the other. Randomness between systems, and within a system, can thwart would-be attackers from doing vulnerability mapping attacks that apply to many machines, or to one machine over a long period of time. Past research has also shown how randomization in algorithms can improve its performance. If we can combine these benefits by applying randomization theory in the creative design of new systems that have built-in moving target defenses, we can indeed achieve the previously conflicting goals of improving security and improving performance at the same time. In the Hardware Security Examples section, I show an example of applying the moving target defense strategy to design hardware cache architectures that improve both the performance and the security of cache memories.

Another approach is to use biologically-inspired defenses, where diversity helps prevent the extinction of a species due to rapidly-spreading viruses and diseases. Concepts of diversity, randomness, and moving target defenses could all be explored, as well as their potential and real impact on improving cyber security.

REWARDING RESPONSIBLE BEHAVIOR

The goal in this research thrust is to realign cyber economic incentives for promoting socially responsible behavior in cyberspace and for deterring malicious behavior.

Today, cyber crime pays: small investments of money and time can yield a large return on investment (ROI). How can we turn the tables so that cyber crime does not pay—or at least does not pay so well? How can we reward responsible behavior in cyber space that may prevent someone else from being attacked, even if we ourselves are not affected? It is said that fear and avarice are the greatest motivators. Since we do not want to wait for a "cyber 911" attack (fear) to trigger improvements in cyber security, can we use economic incentives (avarice) instead to strongly encourage software, hardware, and network vendors and cloud computing providers to build more secure products using the best security design principles and practices? Also, what economic incentives will persuade individual users, companies, and organizations to buy—and even demand—secure computers and services, and to implement security best practices?

Improving cyber security is a bit like improving the environment—although no single entity is ultimately responsible, everyone should be made aware of the consequences and should do their part. For example, the DDoS attacks discussed earlier in this chapter can be significantly mitigated if individual users and corporations do their best to prevent their machines from being used as zombies or bots in such attacks. It may be possible for fines to be levied on corporations with zombie machines, or more generally, for machines where the "security health" metrics are unacceptable. Exhortation alone may not work; note that legislation was needed to enforce car seat use to protect infants in car accidents. However, rather than punitive legislation, the spirit of this research thrust is more toward rewarding responsible behavior. What legislation could provide economic incentives for responsible cyber security behavior, comparable to incentives for using alternative energy sources for conserving energy in "green" buildings and automobiles? But even without any legislation, if the general public is made aware of the security risks, their costs and their potentially disastrous outcomes, they may provide the market pull for vendors to provide more secure IT (Information Technology) products and systems as competitive advantages.

In addition to economic incentives, research into other incentives for responsible cyber security practices is also highly encouraged. This may include the incentive to protect one's reputation in cyberspace. Fear of losing future business if one's reputation is damaged seems to have worked well in promoting responsible behavior when selling goods through web-based sites like eBay.

Hardware Security Examples

General strategies are good, but it is always helpful to see some actual examples. Also, since hardware-enhanced security solutions are less familiar to the general public, and even to the security community, it may be interesting to see some hardware feasibility examples using these general research strategies.

Hardware security mechanisms can satisfy one of the fundamental NCLY cyber security goals—it can significantly increase the work factor for attackers, since it is often orders of magnitude harder to launch a successful attack on hardware than on software. Hence in this section, I discuss the Bastion and Newcache architectures, where hardware security features can be built into future commodity computing devices to enable the construction of more secure computer systems. The Bastion architecture provides hardware and software mechanisms to facilitate the creation of tailored trustworthy spaces, while the Newcache architecture uses a moving target defense strategy to provide more trustworthy hardware that cannot be used to leak confidential information.

ARCHITECTURAL SUPPORT FOR TAILORED TRUSTWORTHY SPACES

Software virtualization technology can be used to protect a trusted application from untrusted applications, and is the core technology underlying cloud computing. Software virtualization creates virtual machines (VMs), which run on a system software layer called a virtual machine monitor (VMM), also called a hypervisor. This hypervisor manages all the hardware resources and isolates one virtual machine from another, giving each the illusion that it has the machine to itself. Trusted virtual machines can be used to run security-critical applications, while untrusted virtual machines run untrusted applications. This allows existing programs to run as normal, using a commodity (untrusted) operating system in the untrusted VM. Meanwhile, trusted applications are isolated by the hypervisor in a new trusted VM.

While software virtualization can provide adequate security for many applications and use cases, it has some unresolved security and performance issues. First, it requires that the entire software stack, including the operating system, be trusted in the trusted virtual machine. However, commodity OSes that are publicly available today with the functionality demanded by applications, are not trusted. While there are some verified OS microkernels like seL4 [4], or ones designed to be trusted like seLinux [5], they are not widely deployed. Second, performance is degraded as costly *world switches* are needed to switch from an untrusted virtual machine to the hypervisor to the trusted virtual machine and back. Third, the isolation of virtual machines executing on the same physical machine provided by software virtualization cannot prevent the leaking of confidential information through the shared hardware resources (e.g., through cache-based side-channel attacks, discussed in the "Mitigating Hardware Information Leaks" section later in this chapter). This can leak secret encryption keys with correctly functional hardware resources, thus defeating any cryptographic protection for confidentiality or integrity.

Furthermore, typically only a small part of an application is security-critical. Rather than trying to certify that the entire large application is trusted, it may be easier to verify only the security-critical modules, i.e., verify that they perform the required functions correctly, do not leak confidential information, and do not contain bugs (security vulnerabilities) that can be exploited by attackers. These security-critical modules are typically much smaller in size than the entire application, and therefore amenable to advances made in static code analysis techniques and theorem proving techniques used to verify their trustworthiness [4]. Frequently, these security-critical modules must run in the same virtual address space as the untrusted application as, for example, in a security monitor running alongside the untrusted application it is monitoring. Hence, a virtual machine may be too coarse a granularity for a trustworthy space; we may need a finer granularity trustworthy space *within a virtual machine.* Possible hardware solutions to these problems are discussed below.

Bastion Security Architecture

Bastion is a hardware-software architecture [3, 6] that can provide secure execution environments for executing trusted software modules in an untrusted software stack. The trusted software modules encapsulate security-critical tasks and must be used to access protected data. In a Bastion system, the operating system in a virtual machine need not be trusted; only the microprocessor and the hypervisor must be trusted in order to enforce the security policies for confidentiality and integrity, for the trusted software modules and their data.

Figure 2 shows a block diagram of the Bastion architecture in a typical virtualized environment. It shows a computer with a hypervisor managing the hardware resources for two virtual machines: one running a Windows OS and the other running a Linux OS. Bastion leverages this virtualized environment, since it is very common in both client computers and cloud computing servers today. Rather than expect the entire hardware platform to be trusted, Bastion only requires that the microprocessor chip be trusted. In particular, the main memory is not trusted (unlike the assumptions for the systems using the Trusted Platform Module (TPM) chip [7], where the entire hardware box is considered trusted, including the main memory and buses). Similarly, rather than requiring the entire software stack to be trusted, Bastion only requires the hypervisor to be trusted. In particular, the guest OS in the VM is not required to be trusted to run the trusted software modules A, B, and C (shown in gray) within untrusted applications on an untrusted commodity OS.

For example, module A could be a security monitor running inside the virtual address space of Application 1, which it is monitoring for security breaches. Clearly, it is important to protect the security monitor itself— that is, protect module A from an attacker exploiting some security vulnerability in Application 1, some other application, or the OS to infiltrate the

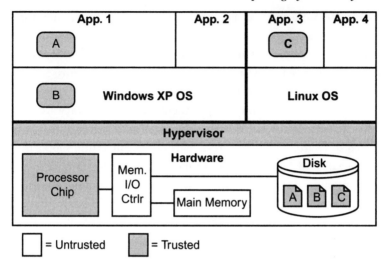

Figure 2. Bastion hardware-software security architecture.

system and attack module A. In general, an application writer may be very motivated to write a secure application (or secure modules within a large application) to protect his security-critical data, but he has no control over other applications or the OS. While OS vendors are highly motivated to secure their OS, it is such a large and complex piece of code that this is an extremely difficult task. Frequent security updates are evidence of the continuing vulnerability of an OS to attacks, despite the best efforts of OS providers. Bastion thus does not require the OS to be bug-free, but instead protects trusted software modules (e.g., A, B, and C) from a compromised OS.

Bastion architecture also provides module A with its own secure storage (shown in Figure 2 as a grey component A, in the local disk or remote storage). This is not a fixed contiguous portion of the disk or online storage, but rather the secure storage consists of any portions of the regular storage medium that are cryptographically secured (i.e., encrypted for confidentiality and hashed for integrity). Only module A has access to the keys to decrypt and verify the hash of its secure storage A. These keys are protected by the trusted hypervisor, which is itself directly protected by the trusted microprocessor in Bastion. This secure storage is persistent (a nonvolatile memory resource) in that it survives power on-off cycles.

Secure and authenticated memory is also provided to module A during its execution. This is volatile memory that does not survive power on-off cycles, but must be protected in order to provide a secure execution environment for a trusted software module like A, B or C in Figure 2. This and other hardware mechanisms [3, 6] provide a fine-grained, dynamically instantiated, secure execution compartment for trusted software. Hence, Bastion can be used to enable tailored trustworthy spaces within a sea of untrusted software, with hardware-hypervisor architectural support.

Figure 2 also illustrates that Bastion can support any number of trusted software modules from many different trust domains simultaneously. For example, Bastion supports trusted software modules in either application space (modules A and C) or OS space (module B). Module B can also have its own secure storage B, which is not accessible to the rest of the OS or any other software. Module C is another trusted software module in a different virtual machine.

Bastion also provides *trustworthy tailored attestation* [3], which enables a remote party to query the state of the trusted components it requires to perform a security-critical task. This can provide an unforgeable report of the integrity measurements of the trusted software modules required for the security-critical task, and of the hypervisor. Input parameters, configurations, and output results of a secure computation can also be included in the attestation report. Attestations can be requested any time, and are very fast.

Minimal Trust Chains and Secure Trust Anchors

An important difference with conventional TPM-based systems [7] is the minimal trust chain used by Bastion, which can skip layers of untrusted software, like the OS and some middleware. Bastion only requires that the microprocessor, the hypervisor, and the relevant trusted software modules form a secure trust chain, skipping layers of untrusted software in between. This minimal trust chain is illustrated on the bottom in Figure 3, while the traditional full trust chain where every layer of software must be trusted and unmodified is shown on the top. Minimal trust chains can enable more resilient execution of security-critical tasks, since they will not be prevented from executing due to updates or malware in unrelated software.

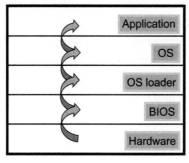

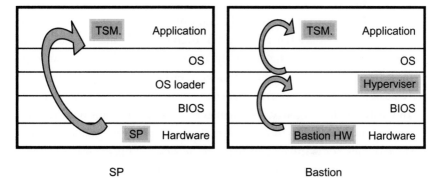

Figure 3. Conventional and minimal trust chains.

However, it is the responsibility of the applications designer to ensure that all the needs of the security-critical task are encapsulated in the trusted software modules and the hypervisor. The integrity of this minimal chain of trusted hardware and software components can then be reported upon a *tailored* attestation request.

While not every software layer has to be verified and attested to in a minimal trust chain, it is important that this chain is securely rooted. In Bastion, this is securely rooted in the trusted hypervisor and the new hardware registers and mechanisms used to protect the trusted hypervisor.

The concept of minimal, layer-skipping trust chains was first proposed in an earlier secure processor architecture called the Secret Protection

(SP) architecture [8, 9]. Here, a trusted hypervisor is not needed, and the untrusted OS runs directly on hardware. SP uses only two new registers as hardware trust anchors to anchor a minimal trust chain for protecting a trusted software module, as shown in Figure 3. It is a simpler security architecture than Bastion, useful for embedded or closed systems, but is not scalable as Bastion for simultaneously protecting multiple, mutually-suspicious trust domains.

MITIGATING HARDWARE INFORMATION LEAKS

New hardware architecture can be used to enhance the security provided by software. For example, hardware can assist software in providing tailored trustworthy spaces and secure trust anchors, as illustrated above by the Bastion and SP architectures. In addition, hardware itself should also be designed to be more trustworthy. For example, it should not leak secret information when operating correctly. Below, I discuss how this can happen, and show an example where a moving target defense is used in hardware design, to design leak-free caches that also improve performance.

In hardware design today, a fundamental strategy used for performance optimization is what I will characterize as a strategy to "make frequent paths fast, but allow infrequent paths to be slow." This enables the average execution time taken to be much closer to the fast path. For example, the frequently referenced memory items are stored in a fast cache memory that is much faster but smaller than the main memory. If the processor wants an item from memory, it gets this item very quickly if the item is found in the fast cache (called *a cache hit*). Otherwise, a *cache miss* occurs and it takes a long time (currently hundreds of processor cycles) to fetch the item from the main memory. This simple strategy has worked well for hardware performance optimizations, reducing the execution time taken.

Unfortunately, the two possibilities of a fast path and a slow path, which can be observed by anyone, or any program, that can time such operations, can be used to encode a binary "0" or "1." Hence, such performance optimization hardware mechanisms can be used to leak secret information through both covert channels and side channels. In a *covert channel attack*, an authorized insider leaks secret information to an unauthorized recipient using mechanisms not intended to be communications channels. In a *side-channel attack*, an insider is not even needed—for example, correctly functioning hardware can leak information without violating any usage or security policies.

In a power analysis side-channel attack [10], for example, an attacker can measure the power consumption of a program, and thus infer secret information such as a secret cryptographic key embedded inside the computing device. While most hardware side-channel attacks require physical access or proximity to the device, cache-based side-channel attacks do not. They can be triggered remotely, and they also do not need any special equipment to obtain the side-channel information [11, 12, 13, 14]. For example, a simple software side-channel attack using the cache can be constructed easily by having a "listener" program run on the same computer as a victim program, accessing data in a large data structure that uses up all the cache lines in the shared hardware cache, and timing each access. When the victim program is scheduled to run, it will replace some of these cache lines with the data it needs. When the listener program is scheduled to run again, it will access its large data structure again and time each access: a fast cache hit indicates his data has not been replaced in the cache, but a slow cache miss indicates that it has been replaced—most likely by the victim program's data. This can be used to leak information about which memory locations the victim program has accessed, since all computers use a static, fixed, memory-to-cache mapping for all programs. From this, the attacker can deduce which table entries were used by the victim program. If the victim program was an encryption program like the Advanced Encryption Standard (AES) [15], the attacker can then deduce which AES table entries were accessed in the cipher and hence what key bits were used to index these tables. This can be used to infer some or all of the bits of the secret encryption key.

Such cache-based side-channel attacks can undermine the strong cryptography used to encrypt secret information to protect its confidentiality. They can also undermine the isolation of different virtual machines provided by hypervisors. In general, the attacker's strategy is simple: exploit shared hardware performance optimization mechanisms like caches or branch prediction mechanisms to leak information. Hardware power optimization mechanisms can also be exploited similarly. Hence, we recommend that new strategies for performance optimization should be researched that can improve performance without sacrificing security.

Newcache: Moving Target Defense in Hardware

The Newcache architecture uses a new moving target strategy in hardware cache design to achieve both performance and security improvements simultaneously [16]. In particular, rather than use the conventional fixed, static mapping of memory addresses to cache lines, it uses a dynamic, randomized mapping of memory addresses to cache lines. Hence, even if the same program is executed on the same computer, it will have a different memory-to-cache mapping each time it executes. Also, the same program executing on a different machine of identical configuration will have a different and unpredictable memory-to-cache mapping. Hence, even if the attacker can observe which physical cache lines hit or miss, he or she cannot deduce any information about the memory locations actually used by a victim program. This thwarts these fast and dangerous cache side-channel attacks—without requiring any software changes in the application programs, nor any changes in the compiler or the instruction set architecture (ISA) of the microprocessor executing the programs. Hence, it is a hardware solution for a hardware information leakage vulnerability. It is also an example that shows that hardware can itself be designed to be more secure and trustworthy; in this case, to not leak information.

In the past, design for security has been at odds with design for performance. One had to choose either a secure system or a high performance system. The Newcache example shows that this need not be the case if we are willing to rethink some basic aspects of computer design. Here, permitting the freedom to do a *clean-slate* design is needed, although the resulting design may in fact not need to change too much of established designs. In Newcache, we only need to change the address decoder—something that has not been changed in the multitude of cache optimization designs proposed in the last three to four decades.

Newcache architecture also shows that using randomization in hardware design can in fact improve both security and performance. This is a surprising result, since cache performance is actually improved due to rethinking the design of caches for security [16]. It is especially surprising in such a mature area as cache design, which has been thoroughly studied by both researchers in academia and practitioners in industry, since caches are perhaps the most important component in the performance of modern computer systems.

Conclusion

With the growing complexity of distributed computer systems and also of computing devices like smartphones, it is impractical to assume that we can achieve perfectly secure systems. Rather, tomorrow's systems must operate securely in the presence of vulnerabilities and malware in the system. They should be resilient and should provide availability of systems and services to security-critical tasks, even under attack.

This chapter discusses some promising game-changing strategies for improving cyber security: enabling tailored trustworthy spaces, thwarting attackers with proactive moving target strategies, and rewarding responsible behavior in cyberspace with economic or other incentives. It is important for academia, industry, and government to work together to research, develop, and incentivize the ubiquitous use of more secure and trustworthy computer products.

Some concrete examples are given to show that these strategies can be used in hardware design to significantly improve cyber security: by either using hardware to enhance the security provided by software, or improving the trustworthiness of the hardware itself. For example, the Bastion hardware-software architecture can be used to help software systems achieve tailored trustworthy spaces, while the Newcache architecture shows how to build more trustworthy hardware—in this case, secure caches that cannot be used to leak information to adversaries—by using a moving target defense in hardware design. Surprisingly, Newcache can improve performance and power-efficiency at the same time as it improves security.

These hardware feasibility examples demonstrate that it is possible to build new hardware features into future commodity processors to enhance cyber security. They are research architectures that should be further evaluated for their validity in different application domains, their robustness against new attacks, and the impact they will have on the current software and hardware ecosystems. Industrial-strength design validation and security verification by teams of professionals are needed before deployment.

The good news is that it may be possible to build security into the hardware and software of future commodity computing and communication devices that can be used to improve cyber security significantly, without degrading their performance or versatility.

REFERENCES

[1] "National Cyber Leap Year Summit 2009 Co-Chairs' Report, Sept 16 2009" (2009, September 16) [Online]. Available: http://www.cyber.st.dhs.gov/docs /National_Cyber_Leap_Year_Summit_2009_Co-Chairs_Report.pdf

[2] Federal Cybersecurity Game-Change R&D Themes (2010, May 19). Kick-off Event 2010 held in conjunction with Security and Privacy Symposium, Oakland, California [Online]. Available: http://www.nitrd.gov/CSThemes /player.html   http://www.nitrd.gov/fileupload/files/CSIAIWGCybersecu-rityGameChangeRDRecommendations20100513.pdf

[3] D. Champagne (2010, April). "Scalable Security Architecture for Trusted Software," PhD Thesis, Electrical Engineering Department, Princeton University [Online]. Available: http://palms.ee.princeton.edu/publications

[4] G. Klein, K. Elphinstone, G. Heiser, J. Andronick, D. Cock, P. Derrin, D. Elkaduwe, K. Engelhardt, R. Kolanski, M. Norrish, T. Sewell, H. Tuch, and S. Winwood, "seL4: Formal verification of an OS kernel," In *Symposium on Operating Systems Principles (SOSP)*, pp. 207–220, October 2009.

[5] seLinux. Web site and papers [Online]. Available: http://www.nsa.gov /research/selinux/

[6] D. Champagne and R. B. Lee, "Scalable Architecture for Trusted Software," Proceedings of the IEEE International Symposium on High Performance Computer Architecture (HPCA), January 2010.

[7] Trusted Computing Group, *TPM Main Specifications* [Online]. Available: http://www.trustedcomputinggroup.org/resources/tpm_main_specification

[8] J. Dwoskin and R. B. Lee, "Hardware-rooted Trust for Secure Key Management and Transient Trust," *ACM Conference on Computer and Communications Security (CCS)*, pp. 389–400, October 2007.

[9] R. B. Lee, P. Kwan, J. P. McGregor, J. Dwoskin, Z. Wang, "Architecture for Protecting Critical Secrets in Microprocessors," *Proceedings of the International Symposium on Computer Architecture (ISCA)*, pp. 2–13, June 2005.

[10] C. Kocher, J. Jaffe, and B. Jun. "Differential power analysis," *Advances in Cryptology – CRYPTO'99*, Vol. 1666 of Lecture Notes in Computer Science, pp. 388–397, 1999.

[11] D. J. Bernstein, "Cache-timing Attacks on AES" [Online]. Available: http://cr.yp.to/antiforgery/cachetiming-20050414.pdf

[12] C. Percival, "Cache Missing for Fun and Profit" [Online]. Available: http://www.daemonology.net/papers/htt.pdf

[13] D. A. Osvik, A. Shamir, and E. Tromer, "Cache attacks and Countermeasures: the Case of AES," *Cryptology ePrint Archive*, Report 2005/271, 2005.

[14] Z. Wang and R. B. Lee, "New Cache Designs for Thwarting Software Cache-based Side Channel Attacks," *Proceedings of the 34th International Symposium on Computer Architecture (ISCA 2007)*, pp. 494–505, June 2007.

[15] NIST, "FIPS-197: Advanced Encryption Standard," November 2001.

[16] Z. Wang and R. B. Lee, "A Novel Cache Architecture with Enhanced Performance and Security," *Proceedings of the 41st. Annual IEEE/ACM International Symposium on Microarchitecture (Micro)*, pp. 88–93, December 2008.

Practical Vulnerabilities of the Tor Anonymity Network

Paul Syverson

Onion routing is a technology designed at the U.S. Naval Research Laboratory to protect the security and privacy of network communications. In particular, Tor, the current widely-used onion routing system, was originally designed to protect intelligence gathering from open sources and to otherwise protect military communications over insecure or public networks, but it is also used by human rights workers, law enforcement officers, abuse victims, ordinary citizens, corporations, journalists, and others. In this chapter the focus is less on what Tor currently does for its various users and more on what it does not do. Because Tor is used at such a high level for law enforcement and national security applications, it faces more significant adversaries than most other uses. I discuss some of the types of threats against which Tor currently offers only limited protection and the impacts of these on all classes of users, but especially on those most likely to confront them.

We have designed and built the Tor anonymity network [3] to secure cyberspace and empower cybercitizens. It is thus squarely in the midst of this volume's concerns. But in law enforcement, often the first thought that comes to mind when one says "anonymity" is of a roadblock against pursuing the source of an attack or other crime. Although this is sometimes the first image that comes to mind, it is not generally the first encounter law enforcers have with anonymity. Typically law enforcers themselves have begun using anonymity long before they observe any criminal activity, and they may even use it to prevent a crime from occurring at all.

As a simple, mundane example of anonymity technology used by law enforcers, consider unmarked vehicles. These are used precisely to avoid obvious distinguishability from other cars around them. This might be to avoid alerting a criminal to the presence or location of law enforcement, or to help protect someone being discretely transported, or for various other reasons. Anonymity is an essential part of law enforcement. Note that unmarked cars are effective, not just because unmarked law-enforcement vehicles are not immediately identifiable as law-enforcement vehicles, but also because most vehicles used by others are similarly anonymous. You can't be anonymous by yourself, and the anonymity protection on which law enforcement depends only works if others have it as well. I return to this point below. Unmarked vehicles are of course just one example. The same applies equally to crime prevention programs, witness protection, anonymous tip lines, and so on. Although there are many important anonymity technologies, my focus in this chapter is on anonymous Internet communication.

Tor protects your anonymity by bouncing your Internet traffic over an unpredictable route comprised of volunteer-run traffic relays all over the world. Tor builds a cryptographic circuit over three relays, and the cryptography prevents each relay from knowing about the other parts of the circuit it does not talk to directly. Only the Tor software on your computer knows about all three relays. Early versions of onion routing laid the cryptographic circuit using an onion, a data structure comprised entirely of layers (one for each hop in the circuit) with effectively nothing at the center. This is what gave onion routing—and Tor (Tor's onion routing)—its name, although circuits in Tor are built in a slightly different way for improved security. What is not different is that Tor gets its protection from the difficulty an adversary has observing the whole circuit. If you are browsing the web over Tor, an adversary might see some of your encrypted traffic go by if he can watch it enter the first relay in the circuit. Similarly,

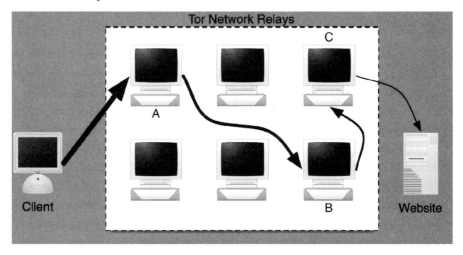

Figure 1. Web browsing through a Tor circuit.

he might see some encrypted traffic headed to you if he can watch the server's response enter the last relay in the Tor circuit. But, unless he can watch both places at once, he will not be able to associate you with the website you are visiting. A basic picture of a Tor circuit is shown in Figure 1.

Unfortunately, the bad guys have a significant advantage over the good guys when it comes to keeping their communications anonymous. For example, if they want to make phone calls with guaranteed anonymity, they can just steal a cell phone off of a restaurant table and then toss it when done. This avenue is not available to those unwilling to break the law. Similarly for Internet communications, bad guys have easy access to cheap and plentiful compromised computers on botnets. As a bonus, their communication can masquerade as not even attempting to appear anonymous while misdirecting attention to the victims of a botnet compromise.

Networks like Tor do not provide this kind of anonymity, but they can help secure the communications of cybercitizens and those who strive to protect them. Because the public Tor network is comprised of thousands of relays in the network all over the world [5], it would be difficult for an adversary to watch all, or even most, of them. This protects ordinary users from revealing a profile of their online interests to common identity thieves, for example, but it is a computer security mechanism, not magic. And like all computer security mechanisms, Tor is most effective if those who use it understand what protections they have and what protections they do not have.

In "Tor: Uses and Limitations of Online Anonymity" in this volume, Andrew Lewman describes how Tor empowers law enforcers and many kinds of cybercitizens, and how Tor makes the Net a safer place by protecting them against traffic analysis. In this chapter I talk about some areas where protecting communications against traffic analysis still has a ways to go.

The Ends of Watching the Ends

Onion routing cryptographically changes the appearance of the traffic as it passes along the circuit. But some features of the communication are still apparent to an observer. Consequently, if an adversary does not see both ends of a Tor circuit, anonymity is typically preserved. If an adversary can see both ends of a Tor circuit, he can trivially correlate who is talking to whom over that circuit. This is thus generally known as an end-to-end correlation attack.

There are many ways to correlate the end points of Tor communication. Since Tor is used for things like web surfing and chatting, communication must be low-latency: anything entering the circuit at one end needs to pop out at the other end pretty quickly. Nobody wants to wait several seconds, let alone several minutes, for a web page to load or to receive the next line in a chat. Besides providing an unacceptably bad user experience, communication protocols for these applications often simply fail with delays of that magnitude.

This means that an adversary who can see both ends of the Tor circuit— for example, because he controls both the first and last relay in the circuit— can simply see the same unique pattern of communication popping out of one end of the circuit that he saw entering at the other end. Just passively watching has long been shown to be adequate to guarantee identification with no false positives [18]. In fact, it is not even necessary to wait for, say, a web request and response to flow over the Tor circuit; just watching the Tor circuit setup is enough [1]. There are many ways to do correlation. Instead of timing, the adversary can monitor volume by counting the traffic that has passed over the circuits of each relay he is watching [20].

Why not just pad all the traffic between the user and, say, the second relay in the Tor circuit? That way all circuits will look the same to those seeing their first relays. Padding and related proposals may have some effectiveness when the applications permit significant latency, such as email. But they have been extensively explored and have significant problems. First, all circuits that are supposed to be indistinguishable must be padded to the same level. This means predicting how much traffic will go over them or delaying traffic that exceeds the maximum rate of flow. Padding everything in this way would mean a large additional load on a public shared network. Also, it is not enough to make the flows all the same. The circuits must all begin and end at exactly the same time.

But aside from the difficulty and the excessive overhead, this will not work anyway if the adversary is able to be at all active. For example, he can corrupt the traffic passing through a relay and then watch for corrupted traffic elsewhere in the circuit. This is sometimes called *bit stomping* or a *tagging attack* [2]. Vulnerability to such attacks has been a recognized limitation of onion routing since its origins in the mid-nineties [9]. If he wants to be more subtle still, the adversary does not have to corrupt the bits, just delay them a little and watch for the delay pattern at the other end. Research has shown that an adversary can add recognizable patterns to low-latency anonymous communication that collaborators will still be able to read even if the traffic is perturbed between them, but that will be undetectable by others [22, 23].

There is at least one system design for resisting even active attacks [8]. This work has enough plausibility to merit further exploration. At this time, however, it seems unlikely that the practical overhead and usability issues can ever be adequately resolved. Even if they are, however, there is another problem.

Onion routing does not get its security from everyone being indistinguishable, even though the attacker is seeing all of their communications. In this, it differs from most anonymous communication models. To achieve that indistinguishability, it is necessary that the attacker not recognize the same messages even when he sees them elsewhere in the system and even if they are encrypted. And that requires synchronization of behavior by all who would be indistinguishable. For the above reasons, among others, this means that the set of individuals whose communication is indistinguishable cannot grow to the scale of Tor's hundreds of thousands of users. So even if they are hiding among each other better, they are hiding in a much smaller *anonymity set.*

Practical systems for anonymous communication based on trying to achieve indistinguishability do exist [14, 2]. Besides being high-latency and thus much less usable, they have never had more than a few hundred concurrent users hidden by a network of a dozen or so relays. This is fine if the goal is simply plausible deniability or for a limited application setting such as municipal voting. It is also fine if the adversary is only watching locally— for example, if he is watching your wireless connection at a coffee shop and nothing else. But in intelligence gathering and law enforcement, this is typically inadequate. If the adversary has the incentives and resources of a nation-state or of organized crime, then it is significant if he knows that, for example, one of a hundred participants is definitely from law enforcement. Also the small anonymity set means that it is now within the resource constraints of the adversary to closely scrutinize the online and offline behavior of everyone identified as participating—at least for the kind of adversary faced in law enforcement and national security settings, so he can learn which participants were from which law enforcement organization even if this was not immediately identified by the communication. In this way, systems designed to resist global observers and active attackers are actually less secure than Tor [21].

End-to-end correlation is not the only way to attack Tor users, but I will not get into the others. (But see for example [15, 16, 6, 11].) Though some attack types have been shown to be implementable, they are mostly much more obscure and less practical than the kinds of attacks we have already described. More importantly, they generally require a lot more groundwork, other assumptions about the network, or the success of different concurrent attacks, not to mention a technically sophisticated adversary. Compared to these, it requires very little technical sophistication to conduct the correlation attacks I have mentioned. Correlation attacks are also generally easy to scale with the resources of even an unsophisticated adversary: the more resources, the more relays he watches. Additionally, since relays are run by volunteers, he can run some himself.

There is another attack that an adversary can do. It is not nearly as accurate as correlation, and it requires some planning and prediction. But it does not require much more sophistication than correlation. Suppose an adversary goes to a website such as www.navy.mil or www.cnn.com and records the pattern of bits that flow back and forth as he downloads the home page or possibly even just the totals. If he later sees a Tor client exchanging the same pattern over a circuit, even though it is encrypted he can have some indication of what website the client is visiting. This is called *website fingerprinting* [10]. It assumes that the website fingerprint is sufficiently unique and has not changed since the adversary visited (or at least that the patterns and sizes have not changed even if the specific content has). It also assumes that the adversary already has the website fingerprint in his database. But if those are all true, then he only has to watch one end of the connection. For example, as mentioned before, he could eavesdrop on your WiFi connection at a coffee shop. Tor already does much better against website fingerprinting than other web anonymizers because it sends all communication in uniform-size chunks [13].

Still, if an adversary is aware of one or more websites of interest that are relatively unique, he can use website fingerprinting as an indicator of those users that he might want to scrutinize further. And, if he is sophisticated, he can also use machine learning techniques to improve his accuracy. Fortunately for the aware user, it is easy to defeat website fingerprinting. By simultaneously visiting multiple sites over the same Tor circuit (which Tor allows automatically) any individual fingerprint becomes hard to recognize [19].

Link Attackers

Most adversaries are not in a position to be directly sitting on a significant fraction of the Tor relays. But does an adversary even need to do that to watch both ends of a Tor connection? No, as it turns out. An adversary can actually observe from a small number of locations and still see much of the traffic on the Tor network.

Figure 1 is a fairly typical picture of communication over a Tor network—a graph of clients, network relays, and destinations with arrows representing the links between them. What such pictures ignore is the structure of the Internet that routes the traffic along those links.

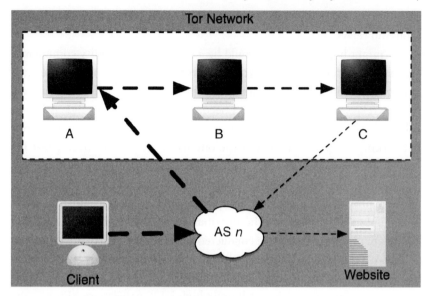

Figure 2. End-to-end correlation by a single AS observer.

The Internet is composed of thousands of independent networks of various sizes called *autonomous systems* (ASes). As traffic moves between a Tor client and a relay in the Tor network, it typically traverses multiple ASes. We have known for years [7, 17] that if the same AS appears on the path from the client to the anonymity network and from the anonymity network to the client's destination, such as a website, then an observer located at that AS can perform a correlation attack to identify the client and her destination. So it is possible to do a correlation attack from one location rather than requiring two. Figure 2 represents such an attack.

In 2004, Feamster and Dingledine showed that the vulnerability to such an attack on the public Tor network was significant [7]. They looked at several client locations in the United States and several plausible destinations such as the websites of CNN or Google and other destinations where users might want to protect their anonymity. For many of the client-destination pairs the probability that a single AS could observe both sides of the connection was over 50 percent. The mean overall probability was about 38 percent. But they mostly looked at client locations in the United States, and at the time of their study the Tor network only contained thirty-three relays. Since that time the network has grown by between one and two thousand relays.

We might hope that there would be a drop in vulnerability commensurate with the tremendous growth of the network. Intuition from the anonymity literature suggests that as the Tor network grows and volunteers operate relays all over the world, it becomes less likely for a single AS to be able to observe both ends of a connection. Intuition from communications networking is more muddy. On the one hand, there have been great increases in both the size and the geographic diversity of the Tor relay network. On the other hand, this might not be reflected in the number of network providers involved, and corporate and service consolidation could even imply a contraction of the distribution of ASes involved in carrying Tor traffic. In experiments we conducted about four years after the original study, we did find a drop in the probability of a single AS-level observer seeing both ends of the connection. But, it only dropped from about 38 percent to about 22 percent, which is still quite high [4]. Furthermore, this was the average, but the drop was not uniform. For 12.5 percent of the client-destination pairs, the chance of such a vulnerability actually increased.

Another issue is that the original Feamster-Dingledine study was only meant to show that such a vulnerability was realistic for plausible source-destination pairs. They did not try to determine what the distribution of actual traffic on the Tor network was. I also studied this question and found 2,251 client ASes and 4,203 destination ASes. For both client and destination sides of circuits, less than 2 percent of all the ASes we recorded accounted for over half of the connections [4].

Tor already does some things by default to counter the chance of a single AS being able to observe both ends of a circuit. For one thing, Tor requires that the first and last relay cannot have network addresses from the same neighborhood. (It is a bit different than requiring that each relay reside in a different AS, but the effect is roughly comparable.) Simulations showed that there is about an 18 percent chance that a single AS will be in the position to observe any Tor connection as of late 2008. There are things that can be done to reduce this to about 6 percent, with an increase to overall Tor communication and computation that is not trivial, but also not prohibitive.

So far, research vulnerability has been limited to a single AS observer. But there is no reason an adversary will be present at only one AS. It would be expected that a nation-state or large criminal adversary will have a much larger reach among both the Tor relays and the links between them.

Lessons for Law Enforcement and Others

Lesson one is that Tor guards against traffic analysis, not traffic confirmation. If there is reason to suspect that a client is talking to a destination over Tor, it is trivial to confirm this by watching them both. This applies to criminals being investigated, which is welcome news. But it also applies to those trying to protect us from them and to those being protected as well, which is not. As already noted, criminals have stronger mechanisms available to them, but Tor can help level the playing field.

Lesson two is that the current Tor network is not by itself sufficient to protect against all of the significant adversaries that may oppose law enforcement or defenders of national security. Tor is a volunteer-run network. Tor picks routes so that relays in a single circuit will generally be from different network and geographic locations. But nothing prevents a large-scale adversary from breaking into or contributing many relays in many locations. An adversary that controls or observes a significant fraction of the network will observe a significant fraction of the network traffic. I have also ignored so far here the nonuniformity of network relays. Some carry a much larger fraction of traffic than others, which only exacerbates this vulnerability.

Lesson three is an extension of lesson two. For correlation vulnerability, the communication links between the relays matter as much as the relays themselves, if not more. An adversary that can observe even a small number of strategic Internet links can do end-to-end correlation on a large fraction of the traffic that passes over Tor.

Fortunately Tor is good enough for many of us. An abusive ex-spouse may be determined enough to monitor or break into a mutual friend's communication to try to find you, but is still unlikely to be able to track you if you communicate with that friend over Tor. See in this volume Andrew Lewman's "Tor: Uses and Limitations of Online Anonymity" for several other examples from the mundane to the exotic. Things are a bit more difficult, however, for those who oppose more resourceful and well-resourced adversaries.

Conclusion

We need a network like Tor, but robust against the vulnerabilities described in this chapter. Unfortunately, no such network now exists. Fortunately, work on one is in the research stages. By Tor's very nature, there is no way to preclude adversary control of either a large fraction of relays or a significant number of Internet links connecting to them, or both—at least not if it is to continue enjoying and deserving the trust of its users. To resist significant adversaries, we might be tempted to run our traffic over a trusted private network of relays under our control. This may obscure some things, but it will still reveal to our adversaries all the connections between the open Internet and this trusted network (hence to or from us). It also provides adversaries a relatively smaller target that will thus be easier to cover entirely. If, however, we make use of trusted and untrusted relays appropriately, we can get the benefits of trusted relays without the problems from their association to us [18, 12]. For example, we can start a circuit at highly trusted relays, then move on to progressively less trusted, but also less associated relays. Of course we must also consider trust in Internet link elements such as ASes. Research in this area is still in the early stages.

For now, there may be times when stealth operations can be constructed with no overt connection to the organization behind them. But this is not always feasible, and even if it is, defense in depth counsels reducing the risk of this relationship being discovered. It may help to initiate our communication over the Tor network via a collection of relays that we trust, that are part of the larger network, and that we have links to that are not publicly visible.

Especially this may help if we can hide our association with these relays. This may help, but its security implications have not yet been fully analyzed. This approach also will not help when we need to communicate to a trusted but associated destination, such as the home office, starting at an insecure location, such as a hotel room under observation of the adversary.

We hope to have clearer advice about reducing your anonymity vulnerabilities soon, as well as better systems for reducing those vulnerabilities. For now, let us hope that a greater understanding of the vulnerabilities and risks of using Tor as it now exists will help empower the cybercitizens that rely on it.

REFERENCES

[1] K. Bauer, D. McCoy, D. Grunwald, T. Kohno, and D. Sicker. "Low-resource routing attacks against Tor," In Ting Yu, editor, *WPES'07: Proceedings of the 2007 ACM Workshop on Privacy in the Electronic Society*, pp. 11–20. ACM Press, October 2007.

[2] G. Danezis, R. Dingledine, and N. Mathewson. "Mixminion: Design of a type III anonymous remailer protocol," In *Proceedings, 2003 IEEE Symposium on Security and Privacy*, pp. 2–15. Berkeley, CA, May 2003. IEEE Computer Society.

[3] R. Dingledine, N. Mathewson, and P. Syverson. "Tor: The second-generation onion router," In *Proceedings of the 13th USENIX Security Symposium*, pp. 303–319. USENIX Association, August 2004.

[4] M. Edman and P. Syverson. "AS-awareness in Tor path selection," In Somesh Jha, Angelos D. Keromytis, and Hao Chen, editors, *CCS'09: Proceedings of the 16th ACM Conference on Computer and Communications Security*, pp. 380–389. ACM Press, 2009.

[5] K. Loesing et al. (2010, December). Tor metrics portal. [Online]. Available: https://metrics.torproject.org/

[6] N. S. Evans, R. Dingledine, and C. Grothoff. "A practical congestion attack on Tor using long paths," In *Proceedings of the 18th USENIX Security Symposium*, pp. 33–50. Montreal, Canada, August 2009. USENIX Association.

[7] N. Feamster and R. Dingledine. "Location diversity in anonymity networks," In S. De Capitani di Vimercati and P. Syverson, editors, *WPES'04: Proceedings of the 2004 ACM Workshop on Privacy in the Electronic Society*, pp. 66–76. Washington, DC, USA, October 2004. ACM Press.

[8] J. Feigenbaum, A. Johnson, and P. Syverson. "Preventing active timing attacks in low-latency anonymous communication [extended abstract]," In M. J. Attallah and N. J. Hopper, editors, *Privacy Enhancing Technologies: 10th International Symposium, PETS 2010*, pp. 166–183. Springer-Verlag, LNCS 2605, July 2010.

[9] D. M. Goldschlag, M. G. Reed, and P. F. Syverson. "Hiding routing information," In Ross Anderson, editor, *Information Hiding: First International Workshop*, pp. 137–150. Springer-Verlag, LNCS 1174, 1996.

[10] A. Hintz. "Fingerprinting websites using traffic analysis," In R. Dingledine and P. Syverson, editors, *Privacy Enhancing Technologies: Second International Workshop*, PET 2002, pp. 171–178. San Francisco, CA, USA, April 2002. Springer-Verlag, LNCS 2482.

[11] N. Hopper, E. Y. Vasserman, and E. Chan-Tin. "How much anonymity does network latency leak?" *ACM Transactions on Information and System Security*, Vol. 13, No. 2, pp. 13–28. February 2010.

[12] A. Johnson and P. Syverson. "More anonymous onion routing through trust," In *22nd IEEE Computer Security Foundations Symposium, CSF 2009*, pp. 3–12. Port Jefferson, New York, USA, July 2009. IEEE Computer Society.

[13] M. Liberatore and B. N. Levine. "Inferring the source of encrypted HTTP connections," In R. N. Wright, S. De Capitani di Vimercati, and V. Shmatikov, editors, *CCS'06: Proceedings of the 13th ACM Conference on Computer and Communications* Security, pp. 255–263. ACM Press, 2006.

[14] U. Moller, L. Cottrell, P. Palfrader, and L. Sassaman. "Mixmaster protocol—version 3," IETF Internet Draft, 2003.

[15] S. J. Murdoch. "Hot or not: Revealing hidden services by their clock skew," In R. N. Wright, S. De Capitani di Vimercati, and V. Shmatikov, editors, *CCS'06: Proceedings of the 13th ACM Conference on Computer and Communications Security*, pp. 27–36. ACM Press, 2006.

[16] S. J. Murdoch and G. Danezis. "Low-cost traffic analysis of Tor," In *2005 IEEE Symposium on Security and Privacy,(IEEE S&P 2005) Proceedings*, pp. 183–195. IEEE CS, May 2005.

[17] S. J. Murdoch and P. Zieliński. "Sampled traffic analysis by internet-exchange-level adversaries," In Nikita Borisov and Philippe Golle, editors, *Privacy Enhancing Technologies: 7th International Symposium*, PET 2007, pp. 167–183. Springer-Verlag, LNCS 4776, 2007.

[18] L. Øverlier and P. Syverson. "Locating hidden servers," *In 2006 IEEE Symposium on Security and Privacy (S&P 2006)*, Proceedings, pp. 100–114. IEEE CS, May 2006.

[19] L. Pimenidis and D. Herrmann. "Contemporary profiling of web users," *Presentation at the 27th Chaos Communication Congress (27C3)*, December 2010.

[20] A. Serjantov and P. Sewell. "Passive attack analysis for connection-based anonymity systems," In E. Snekkenes and D. Gollmann, editors, *Computer Security—ESORICS 2003, 8th European Symposium on Research in Computer Security*, pp. 141–159, Gjøvik, Norway, October 2003. Springer-Verlag, LNCS 2808.

[21] P. Syverson. "Why I'm not an entropist," In *Seventeenth International Workshop on Security Protocols*. Springer-Verlag, LNCS, 2009. Forthcoming.

[22] X. Wang, S. Chen, and S. Jajodia. "Tracking anonymous peer-to-peer voip calls on the internet," In C. Meadows and P. Syverson, editors, *CCS'05:*

Proceedings of the 12th ACM Conference on Computer and Communications Security, pp. 81–91. ACM Press, November 2005.

[23] W. Yu, X. Fu, S. Graham, D. Xuan, and W. Zhao. "DSSS-based flow marking technique for invisible traceback," In *2007 IEEE Symposium on Secuity and Privacy (SP'07)*, pp. 18–32. IEEE Computer Society, May 2007.

Defending Software Systems against Cyber Attacks throughout Their Lifecycle

Hira Agrawal, Thomas F. Bowen, and Sanjai Narain

Malware usually enters a distributed software system along three avenues. First, it may be hidden surreptitiously within application code by a malicious developer. Examples of malicious code include Trojan horses, backdoors, and logic bombs. This code can be triggered by the developer or his accomplices—*after* the application has been deployed in the field—using secret input values that are known only to them. Current malware detection techniques do not detect such code. Telcordia's Software Visualization and Analysis Toolsuite (TSVAT) system [5] helps detect such code by combining static program analysis and dynamic program testing techniques. TSVAT "forces" the tester to provide the secret trigger value in the form of a test case as early as possible, so its malicious effects can be discovered in a protected test environment, or by examining the code that makes up its trigger conditions and/or actions. TSVAT has been used both internally within Telcordia and externally by its customers. Telcordia is also currently extending this tool for use by the U.S. Army under their CERDEC TITAN Information Assurance program [3, 4].

The second avenue along which malware can enter a software system is via an accidental or a deliberate misconfiguration of the deployment environment. Not just applications but their host operating systems and networks may be misconfigured as well. Configuration is the glue for logically integrating applications to satisfy their end-to-end requirements. Requirements address not only control and data flow between applications but also the security, performance, and reliability of the system as a whole. These requirements are implemented by setting the hundreds of configuration parameters, or *knobs*, to precise values. However, today, the large conceptual gap between requirements and configurations is bridged manually. Thus, a large number of configuration errors arise. It is well-documented that these account for 50 to 80 percent of vulnerabilities and downtime in cyber infrastructure [20, 23]. Examples of configuration errors include incorrect access-control policies that let an adversary access services he shouldn't, and addressing errors that can disconnect network nodes. Telcordia's ConfigAssure system [9, 13, 14, 15] provides fundamental tools for eliminating configuration errors, including a high-level requirement specification language and engines for configuration synthesis, diagnosis, repair, reconfiguration planning, verification, and visualization. These tools are built by leveraging modern SAT-based constraint solvers [12]. They are being deployed in DISA's Multinational Information Sharing network that supports collaboration between the United States and its allies. They have also been successfully trialed on NSA's High Assurance Platform that supports virtualization with multilevel security. These tools are also noninvasive. They operate purely by analysis of system configuration. They do not need to touch system components.

The third avenue allowing malware to enter a software system is post-deployment exploitation of a programming flaw in an application. Example flaws include failure to guard against memory corruption that allows buffer-overflow attacks and failure to validate user input that allows command injection attacks. Telcordia's run-time monitoring technology remediates this avenue by checking an application's observed behavior against its normal behavior and blocking execution if significant anomalies are found. Run-time monitoring considers myriad execution artifacts observable by interposing the application/OS interface (through system calls), the application/file system interface, and the application/network interface. By blocking only offending operations, benign functionality can continue. Run-time monitoring creates anomaly-detection rules by continuous observation and learning of normal behaviors. To compensate for the inherently high false-positive rate usually associated with anomaly detection, correlation and causality analysis are employed. The technique is transparent to applications, requiring neither access to application source code nor application modification. Since it is aimed at detecting malicious behaviors rather than the vulnerabilities underlying those behaviors, it is agnostic to the mechanism by which the malicious code comes to be executed. This technique, which was developed in programs funded by NIST and DARPA, has been shown to be effective at detecting zero-day exploits that use previously unseen malicious code. Taken together, our defensive technologies provide a comprehensive defense against attacks on a distributed software system. As depicted in Figure 1, first, TSVAT is used to detect malware that may have been hidden inside the system by malicious developers before it is deployed in the field. Next, Telcordia's ConfigAssure tool is used to detect and eliminate deliberate or inadvertent errors in the system's configuration upon its deployment in the field, thereby precluding malware relying on such misconfigurations from entering the system. Finally, Telcordia's Run-time Monitoring and Anomaly Detection techniques provide an active, ongoing last line of defense against malware that can enter the system by exploiting any residual vulnerabilities it may contain, after it has been deployed in the field. In summary, attacks before deployment are detected by TSVAT. The deployment configuration is secured by ConfigAssure.

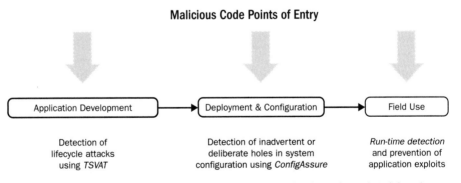

Figure 1. Defending software systems against cyber attacks throughout their lifecycle.

Attacks after deployment are detected and remediated by the run-time monitoring system.

The following section describes the TSVAT system. Then the "Detecting and Eliminating Vulnerabilities during Deployment" section describes the ConfigAssure system. Following that, the section "Detecting and Blocking Malware after Deployment" describes the run-time monitoring system. The chapter's final section discusses the conclusions we draw from the information presented in this chapter.

Detecting Malware Before Deployment

Application sabotage by malicious developers poses a significant and real threat to the security of critical systems. Malware hidden inside an application is characteristically different from other forms of malware in that it may be programmed to provide incorrect results when supplied with specific, unique inputs that are known only to the malicious developer. Such attacks are not detectable using existing anomaly detection techniques because execution of the malicious code may not result in any new, externally visible events such as new system calls or I/O operations. A software engineer tasked with developing a GPS guided weapon system may, for example, hide malicious logic in its onboard software so it returns an undesirable launch trajectory when the supplied target coordinates match the coordinates of certain enemy locations stored within the application. The malicious code, in this case, lies dormant until awakened by one of those secret coordinates. Similarly, a malicious developer implementing an authentication module for a command and control application may implement an additional "feature" causing authentication to be skipped for specific, secret user names and/or passwords. In this case, the lifecycle attack does not require insertion of extra blocks of code that remain dormant until they are activated in the field. Instead, it consists of an extra condition surrounding the authentication code that always succeeds except when supplied with a user name that matches the secret value. Such lifecycle attacks, when hidden inside a critical system, can obviously inflict serious harm on its safety and integrity.

Lifecycle attacks like those mentioned above cannot be detected using static analysis tools because such tools [8] are aimed at finding programming flaws, such as failing to guard against buffer overflows or failure to validate user input, which may be exploited to inject malicious code into a running application. Lifecycle attacks do not rely on such flaws in their host applications. They cannot be detected using traditional, requirements based testing methods either, because normal, expected use cases are unlikely to provide the secret input values required to trigger them.

The state-of-the-art in lifecycle attack detection consists of best practices that involve putting procedural and technical controls in place [7], such as comprehensive background checks of developers, looking for signs of discontent among developers, secure account management and password controls, instituting the principles of least privilege and separation of duties, strict change control, and logging of all activities, among others. These controls are necessary but not sufficient, as malicious developers can carry out their activities using permissions they already possess. One relevant technique in use today for detection of lifecycle attacks requires manual inspection of all program code. Inspection involves a line-by-line examination of all application logic, which, to say the least, is a highly tedious, time consuming, and error prone process.

TSVAT combines static program analysis and dynamic program testing techniques to help detect lifecycle attacks. It steers program analysts toward coming up with test cases that are likely to trigger conditions associated with lifecycle attacks. This process is driven by a control flow analysis technique [1, 2], which enables analysts to achieve "maximum" additional program coverage with each new test they create, which quickly takes them closer toward covering the triggers and actions associated with lifecycle attacks. This, in turn, "forces" the analysts into examining that code much more closely, which is likely to reveal the associated attacks. Failing that, the attacks are revealed when tests are constructed that cause their trigger conditions and actions to be executed in the protected test environment.

The process outlined above is facilitated by an intuitive graphical user interface that provides a color coded picture of the program, where the highest value coverage target stands out prominently from the rest. It highlights all blocks and branches of the program in different colors depicting what they are worth at that time in terms of how much additional coverage they guarantee if they are covered by the next test case. Every time an analyst runs a new test aimed at covering the most valuable block or branch, the tool provides immediate feedback by changing the colors of program elements reflecting the coverage results of that test, and shifts the analyst's attention to the new highest valued block or branch.

If the target program does contain a lifecycle attack, the corresponding trigger branches and action blocks remain uncovered, as they are designed to be exercised under conditions known only to their developers. Thus, once other higher valued blocks and branches have been covered, the focus will quickly shift toward malicious code triggers and actions, if any. When that occurs, the analysts would be forced to examine the conditions under which they would be covered, and the actions that would be taken when those triggers are satisfied. The resulting scrutiny will, very likely, reveal the malicious code. Failing that, when tests that cover those blocks or branches are run, the corresponding attacks would be triggered and their effects would be visible in the controlled test environment, thereby revealing their existence. In this way, it leads program analysts into finding lifecycle attacks in a systematic, cost efficient manner by helping them achieve the most coverage with as few tests as practically possible.

In summary, current tools and techniques are focused mostly on finding inadvertent programming flaws that can be exploited to inject malicious code into those programs after they are deployed. We, on the other hand, provide a technique and a tool to detect malicious code inserted deliberately into an application before it is put into production. This tool iteratively identifies high value segments of application code, so any test that causes such a segment to be executed is automatically guaranteed to cause a large fraction of previously dormant code to also be executed. This ensures that any hidden malicious behavior in such segments is revealed in a protected test environment. The tool does not, however, construct such tests; it is up to the analysts to devise them by examining the corresponding code. We plan to leverage recent advances in automatic test case generation [6, 10] and combine them with automatic identification of high value unexecuted code, to add a targeted test generation facility to this tool.

Detecting and Eliminating Vulnerabilities during Deployment

This section describes Telcordia's ConfigAssure system, illustrated in Figure 2. It provides fundamental tools for eliminating configuration errors that are responsible for 50 to 80 percent of downtime and vulnerabilities in

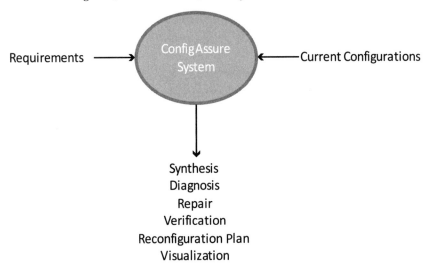

Requirements ⟶ ConfigAssure System ⟵ Current Configurations

Synthesis
Diagnosis
Repair
Verification
Reconfiguration Plan
Visualization

Figure 2. ConfigAssure system overview.

cyber infrastructure. These tools support requirement specification, configuration synthesis, diagnosis, repair, verification, visualization, and reconfiguration planning. The tools are noninvasive in that they work purely off configurations of cyber infrastructure components. They do not need to touch the infrastructure.

Today's configuration languages compel administrators to choose the precise value of each configuration parameter to satisfy requirements. But the choice made to satisfy one requirement may be inconsistent with that for another. Manually removing these inconsistencies is hard because of the large number of requirements and configuration variables and their possible values. ConfigAssure's specification language allows one to directly specify requirements as constraints on configuration parameters. Intuitively, a constraint represents a set of acceptable values.

A constraint solver efficiently computes the intersection of all of these sets to synthesize parameter values satisfying all constraints. Requirements can also be solved in the context of current configurations by conjoining these with the conjunction of constraints of the form $x = c$, where x is a configuration parameter and c is its current value.

If the intersection is empty, a root cause listing values in the current configuration that contribute to unsolvability is computed. This root cause represents a diagnosis. These values are iteratively relaxed until an intersection is computed, in effect repairing these values.

To verify that a property always holds, the property is expressed as a constraint and its negation is checked for *unsolvability*. For example, to verify semantic equivalence of firewall policies, each policy is expressed as a constraint on fields of an IP packet that is true if and only if the policy admits that packet, and then it is shown that it is impossible for one constraint to hold but not the other.

The visualization of a large number of logical structures latent in the configuration is computed. These structures are not just the topologies at different protocol layers but also routing domains and relationships between these. Their visualizations reveal deep structural vulnerabilities such as single points of failures.

Figure 3 shows an example of a visualization of the IP topology latent in the configurations of eight routers. It clearly shows that MEMPHIS is a single point of failure for reachability between MIAMI and the rest of the world. It also shows two disconnected networks on the right. An attacker could exploit these vulnerabilities to isolate MIAMI by launching a denial of service attack against MEMPHIS. He could also attack nodes on the right side of the network without fear of being detected at any node on the left side. The misconfigurations that allow these vulnerabilities can be eliminated by ConfigAssure.

After repair to the current configuration has been computed, the problem of how to safely apply the repairs to components remains. Reconfiguration planning is hard because the space of all possible plans includes all permutations of all configurations. In general, all components cannot be concurrently reconfigured because an out-of-band management network cannot always be assumed. Thus, a reconfiguration plan is computed that defines the order in which to change the configurations so that a safety invariant is never violated at any step. Safety invariants are used for preservation of mission-critical services and security properties. A safety invariant is transformed into a constraint when configurations change. A solution to this constraint represents a reconfiguration plan.

ConfigAssure is implemented with modern SAT-based constraint solvers. It has been developed with funding from IARPA, NSA, DISA, and AFRL, in collaboration with Princeton, MIT, and Penn State. It is being

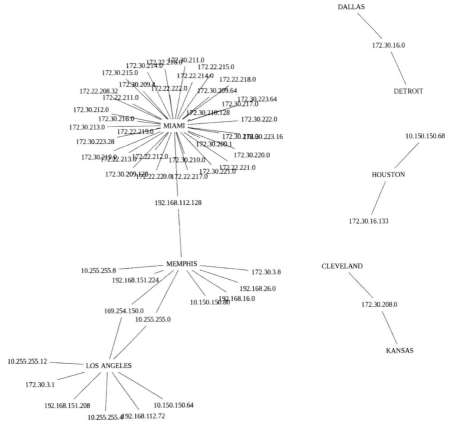

Figure 3. Visualization of IP network topology noninvasively computed from analysis of router configuration files.

deployed at DISA's Multinational Information Sharing infrastructure. This infrastructure allows one to set up logically separated IP virtual private networks. ConfigAssure has also been successfully trialed with NSA's High Assurance Platform. This platform allows one to set up multiple, logically separated virtual machines at different security levels on the same physical machine running the SELinux secure operating system.

Current tools for deployment vulnerability analysis are typically invasive. They face scalability challenges, do not identify configuration errors, and are impractical for assessing vulnerabilities such as a single point of failure. Current policy-based networking tools [19] are procedural and hence do not solve the above fundamental problems. For example, it is impractical to write down repair procedures for each violated requirement. Procedures for different requirements may conflict because the procedures still specify concrete values to repair to.

Detecting and Blocking Malware after Deployment

Despite approaches applied during development, testing, installation, and upgrade, residual security vulnerabilities remain in deployed applications. In this section we present our run-time process monitoring and control solution to detect and remediate exploitation of residual vulnerabilities while preserving unaffected functionality.

Run-time monitoring, as shown in Figure 4, interposes security-relevant process interfaces to expose low-level process behaviors, and applies anomaly detection (AD) style analysis to the behaviors to first build profiles of normal process behavior and then detect deviations as attack indicators. The interposed interfaces are between the process and (1) system calls, (2) the file system (and for Windows OSs, registry), and (3) the network interface.

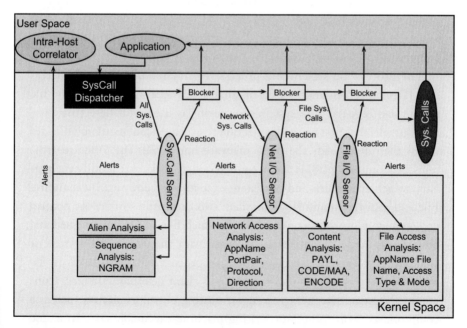

Figure 4. Run-time monitoring and anomaly detection.

Interposition likewise enables reaction to deviations through blocking. If profiles were perfect, individual process monitoring would be sufficient; however, profiles are imperfect since AD is prone to false positives. To compensate, an analysis hierarchy is used. Since most attacks involve multiple processes and computers, the analysis hierarchy performs inter-process and inter-computer correlation—and causality analysis to distinguish between benign anomalies and malicious attacks. Because run-time monitoring is based on AD, it is best suited to enterprises operated under strict controls, such as limiting ad-hoc installation of new applications, imposing rigid application upload controls, and requiring thorough pre-deployment testing. It is less suited for enterprises with less rigid operating controls.

The process/system call monitors support detection [11] and remediation [16] based on analysis of system call sequences. System call monitoring is especially useful for code injection attacks, which inject foreign code into a running application by exploiting vulnerabilities. The foreign code typically uses system calls differently than the native code, so injection and execution usually produce system call anomalies. System calls are analyzed with respect to two criteria: Is the current system call one that the application has ever used, and is the current sequence of system calls consistent with prior sequences? Violation of the first criteria is highly indicative of an attack; however, minor system call sequence permutations are frequently benign.

The process/file system and process/registry monitors support detection [21] and remediation based on analysis of access requests to the file system and registry. File system monitoring is useful for detecting two-stage attacks that first use code injection to execute a malicious downloader which then downloads the main malware into a file. For Window OSs, registry monitoring [18] is useful for detecting attacks that modify the registry to achieve future and persistent execution of downloaded malware. File and registry monitoring considers two dimensions of access: control and content. Control is concerned with which files are being accessed and their assess modes. Content is concerned with the data being written or read. Control analysis can detect unusual process behaviors based on file name—for example, a compromised email client deleting .doc files. Content analysis can detect the reading or writing of abnormal data such as a compromised word processor writing machine-executable code to a .doc file. Content analysis is performed in the context of control analysis, hence content anomalies are detected with respect to specific applications and file names. The details of file content analysis are identical to those of network data analysis, described next.

The process/network monitor supports detection and remediation based on each process's network I/O. Process/network monitoring [22] is useful for detecting attacks triggered by incoming network data as well as attacks that emit network data. Like file monitoring, network monitoring performs analysis of control and content. Control is concerned with the source and destination, size, and protocol of transferred network data. Content is concerned with the payload of the transferred data. Control analysis can detect attacks using backdoor communications, such as unusual ports, or attacks that send and receive data blocks of unusual size (an indicator of buffer exploitation). Control analysis is useful in detecting the blatant *phone home* activity that some malware exhibits. Content analysis can detect attacks that require unusual contents in data blocks—for example, data blocks that contain actual executable machine code (an indicator of an injection attack). Specific content analysis algorithms that have proven useful include histogram byte frequency (PAYL), ratio of binary to ASCII bytes (ENCODE), the likelihood that content contains byte sequences that can be successfully executed (CODE), and the likelihood that any discovered executable code is also malicious (MAA). Content analysis is performed in a specific control context, so that content considered unusual for one application, source, and destination combination may be normal for another combination.

The application-specific nature of the profiles means that each monitor can see an especially potent attack indicator: an event from an application for which there is no profile. Such an event indicates that a previously unknown application is running, or that an existing application is behaving in an exceptionally unusual way. Both are strong evidence of attack.

The monitors described all use interposition to position themselves between a process and a system resource; hence the monitor can block access to the resource if the access is malicious. Monitors block access by returning otherwise legitimate errors. For example, the interposer returns "permission error" in response to a malicious file access. For well-written applications, blocking is handled gracefully, and results in precision reactions that allow the process to continue. For poorly written applications (or injected code) blocking may result in serious consequences, such as crashing, although these consequences are preferable to the blocked malicious behaviors.

The monitors all have a nonzero false positive detection rate. Escalation of detected anomalies that are false positives into reactions that interfere with normal operation is prevented through hierarchical analysis [17]. Monitors classify each anomaly based on the likelihood that it is malicious and its potential consequences, and decide if immediate blocking is required. If immediate blocking cannot be justified, anomalies are reported up a chain of analyzers that evaluate them in combination, looking for the presence or absence of correlation and causality to strengthen or weaken the hypothesis that an attack is occurring. For example, consider a fairly typical attack scenario in which an attacker sends a malware infected gif file to a victim via email, such that when the victim opens the gif file it injects code into the viewer application causing download of a rootkit into a hidden file and modification of the registry so that at next login the rootkit will be installed. When the initial email arrives, the network monitor may detect with low confidence that the gif file contains executable malicious code. When the victim opens the email and displays the gif file, the monitors may detect unusual system calls from the infected viewer, as the injected malware executes, unusual network activity as the infected gif viewer retrieves the rootkit, unusual file access as the infected gif viewer stores the rootkit, and unusual registry access as the infected gif viewer updates the registry to arrange for installation of the rootkit. None of these behaviors in isolation may be sufficient to conclude that an attack is occurring, but when taken together, the evidence is compelling. When an attack is detected by hierarchical analysis, it may be too late for blocking reactions to contain the attack, therefore coarser remediation, such as killing infected processes, and deleting the downloaded file, is required.

In summary, the use of application-specific process monitors using anomaly detection techniques is an effective last line of defense against residual software vulnerabilities. The high false-positive rate inherent in anomaly detection techniques can be ameliorated by higher-level correlation and causality analysis of the anomalies reported by individual monitors.

Conclusion

Malware enters a software system along three avenues: it is hidden surreptitiously within applications by a malicious developer, it is inserted into the system due to an accidental or a deliberate misconfiguration of the deployment environment, and it is injected into a running application by a malicious user by exploiting a programming flaw in the application logic. This chapter described three tools developed at Telcordia for blocking all these avenues: TSVAT, ConfigAssure, and Runtime Monitoring, respectively. The first helps application testers conserve testing resources by guiding them to hidden code. The second helps system administrators in creating vulnerability-free distributed application configuration. The third protects against the exploitation of vulnerabilities not caught by any other technique. These tools have been trialed or are being deployed in real enterprises. Together, they offer a comprehensive defense against attacks on software systems throughout their lifecycle. A good future direction would be to integrate these into a single solution.

REFERENCES

[1] H. Agrawal, "Dominators, Super Blocks, and Program Coverage," *ACM Symposium on Principles of Programming Languages*, pp. 25–34, 1994.
[2] H. Agrawal, "Efficient Coverage Testing Using Global Dominator Graphs," *ACM Workshop on Program Analysis Tools and Engineering*, pp. 11–20, 1999.
[3] H. Agrawal, J. Alberi, L. Bahler, W. Conner, J. Micallef, S. Snyder, and A. Virodov, "Detecting Lifecycle Attacks with Predeployment Program Analysis and Testing," The 27th Army Science Conference, 2010.
[4] H. Agrawal, J. Alberi, L. Bahler, W. Conner, J. Micallef, S. Snyder, and A. Virodov, "Preventing Insider Malware Threats Using Program Analysis Techniques," Military Communications Conference (MILCOM), 2010.
[5] H. Agrawal, J. L. Alberi, S. Ghosh, J. R. Horgan, J. J. Li, S. London, N. Wilde, and W. E. Wong, "Mining System Tests to Aid Software Maintenance," *IEEE Computer*, vol. 31, no. 7, pp. 64–73, 1998.
[6] D. Brumley, C. Hartwig, Z. Liang, J. Newsome, D. Song, and H. Yin, "Automatically Identifying Trigger-based Behavior in Malware," *Botnet Detection: Countering the Largest Security Threat*, Advances in Information Security, eds. W. Lee, C. Wang, and D. Dagon, Springer US, 2008.

[7] D. Cappelli, A. Moore, R. Trzeciak, and T. J. Shimeall, "Common Sense Guide to Prevention and Detection of Insider Threats," Software Engineering Institute, Carnegie Mellon University, 2010.

[8] B. Chess and J. West, *Secure Programming with Static Analysis*, Addison-Wesley, 2007.

[9] Formal Methods in Networking. Graduate seminar course, Computer Science Department, Princeton University. [Online]. Available: http://www.cs.princeton.edu/courses/archive/spring10/cos598D/Introduction.pdf, Spring 2010.

[10] P. Godefroid, M. Y. Levin, and D. Molnar, "Automated Whitebox Fuzz Testing," *Network Distributed Security Symposium (NDSS)*, 2008.

[11] S. Hofmeyr, S. Forrest, and A. Somayaji, "Intrusion Detection Using Sequences of System Calls," *Journal of Computer Security*, vol. 6, pp. 151–180, 1998.

[12] S. Malik and L. Zhang, "Boolean Satisfiability: From Theoretical Hardness to Practical Success," *Proceedings of the Communications of the ACM*, August 2009.

[13] S. Narain, G. Levin, V. Kaul, and S. Malik, "Declarative Infrastructure Configuration Synthesis and Debugging," *Journal of Network Systems and Management*, Special Issue on Security Configuration, eds. Ehab Al-Shaer, Charles Kalmanek, and Felix Wu, 2008.

[14] S. Narain, "Network Configuration Management via Model-Finding," Proceedings *of USENIX Large Installation System Administration (LISA) Conference*, 2005.

[15] S. Narain, R. Talpade, and G. Levin, "Network configuration validation," Chapter in *"Guide to Reliable Internet Services and Applications,"* eds. C. Kalmanek, R. Yang, and S. Misra, Springer, 2010.

[16] Niels Provos, *Improving Host Security with System Call Policies*, Center for Information Technology Integration University of Michigan, provos@citi.umich.edu

[17] P. Ning, Y. Cui, and D. S. Reeves, *Constructing Attack Scenarios through Correlation of Intrusion Alerts*, Department of Computer Science, North Carolina State University.

[18] PAR Government Systems Corp, *Countering Insider Threats—Handling Insider Threats Using Dynamic, Run-Time Forensics*. [Online]. Available: http://www.dtic.mil/cgi-bin/GetTRDoc?Location=U2&doc=GetTRDoc.pdf&AD=ADA473440attached.doc

[19] "Policy-based Networking." [Online]. Available: http://www.research.ibm.com/policy/

[20] "Securing Cyberspace for the 44th Presidency" (2008, December). Center for Strategic and International Studies. [Online]. Available: http://csis.org/files/media/csis/pubs/081208_securingcyberspace_44.pdf

[21] S. J. Stolfo, S. Hershkop, L. H. Bui, R. Ferster, and K. Wang, *Anomaly Detection in Computer Security and an Application to File System Accesses*, Columbia University.

[22] K. Wang and S. J. Stolfo, *Anomalous Payload-based Network Intrusion Detection*, Computer Science Department, Columbia University.

[23] "What's Behind Network Downtime? Proactive Steps to Reduce Human Error and Improve Availability of Networks." [Online]. Available: http://www.juniper.net/solutions/literature/white_papers/200249.pdf

Improve Availability of Networks: Internet Exchange Points and Their Role in Cyberspace

Akio Sugeno

The Internet could not exist without *Internet exchange points* (IXPs). The concept of IXPs was developed in the early 1990s, and IXPs have continued to grow in quantity, location, and size (traffic volume) as the Internet has grown. There are, however, very few books or papers written about IXPs. Knowledge of IXPs has long been confined to industry experts. In this chapter, I provide an overview of IXPs along with their roles in the Internet. The first part of this chapter identifies the architecture of the Internet. The second identifies the concept of peering (a prerequisite for IXPs). The third and final part of the chapter identifies IXPs in greater detail. In the conclusion to the chapter I provide a list of organizations which support and contribute to the Internet.

What Is an Internet Exchange Point?

As the name implies, IXPs exchange something, but what? The answer is that Internet service providers (ISPs) exchange their traffic with each other over IXPs. IXPs provide the mechanisms (physical connections in a data center or carrier hotel) that enable ISPs to exchange traffic easily and cost effectively. In other words, it makes business sense for ISPs to exchange traffic over IXPs. To fully understand this point, it is important to first have an understanding of how the Internet works.

As opposed to a traditional telephone network that is owned and administered by a telephone carrier and is therefore considered to be a single autonomous network, the Internet is not a single autonomous network. No one company or organization owns it, and there is no central organization to administer or monitor it. Simply put, the Internet is a collection of ISPs (i.e., a collection of autonomous networks creating a network of multiple autonomous networks). The 36,047[1] ISPs [1] (or autonomous networks) are connected by many different methods, and together they constitute the Internet. As you can imagine, it is difficult to maintain such a network of multiple autonomous networks because each autonomous system (ISP) has different interests, policies, engineering skills, operation skills, and so on. Therefore, it takes an enormous amount of effort by the Internet community to make the Internet work.

Connectivity Options for ISPs

In the previous section, I noted that all ISPs are connected in some way in the Internet. In order for ISPs to connect and exchange their traffic with each other, today's Internet offers two types of connectivity options for ISPs: *transit* and *peering*. The fundamental difference between transit and peering is the scope of reachability. In short, *transit* provides global reachability (i.e., access to any IP address worldwide) whereas peering provides limited reachability (i.e., access to only a limited subsection of IP addresses). You may wonder why we need both, why having transit only will not

---

1 As of November 2010.

suffice. I explain that later in this chapter, but before explaining transit and peering in detail and why both are necessary, I first define some basic technical terms.

BGP: T²HE GLUE CONNECTING ISPS

Being *connected* means an ISP can send and receive (or exchange) its traffic to and from other ISPs. It does not merely mean simple physical connectivity. In the Internet, there is a common language every ISP speaks. This language, more precisely known as *communication protocol*, is called Border Gateway Protocol (BGP) [2]. BGP provides a mechanism that all ISPs worldwide use to connect to each other to form the Internet. Explaining BGP in detail is beyond the scope of this chapter but I will explain BGP very briefly here.

> **ASN and routes:** Because the Internet is a network of multiple autonomous networks, we need an identifier (or name) for each autonomous network (i.e., a name for each ISP's network). The identifier used is the *autonomous system number* (ASN). A *route* is a collection of addresses. An address is an *IP address*. For instance, a route can be written as 198.32.165.0/24, which represents 254 usable addresses (from 198.32.165.1 to 198.32.165.254).
>
> **Announce or no connection:** BGP *glues* ISPs with ASNs and routes. For example, suppose an ISP (ASN 100) has 1,000 routes. The ISP must *announce* the routes via BGP to other ISPs, effectively saying something like "We, ASN 100, have 1,000 routes such as 198.32.165.0/24." This announcement will be propagated throughout the Internet via BGP.
>
> Once another ISP (ASN 200) has received the announcement and updated its routing table so that it effectively reads "We, ASN 200, know ASN 100 has 1,000 routes such as 198.32.165.0/24," *these two ISPs are connected.* At this point, ISP (ASN 200) can send traffic to ISP (ASN 100).

Now that you understand how connections between ISPs are made, the following sections explore the concept of transit and peering.

Transit and Peering

When two ISPs are connected via transit, they are not equal. The relationship is a commercial one (buyer and seller). The key characteristic of transit service is that all routes (usually called *full routes*) announced by all ISPs (36,047 ISPs) worldwide are provided by the seller to the buyer. Once an ISP has received full routes, it can send and receive traffic to and from anywhere.

When two ISPs are connected via peering, they are basically equal. There is usually no monetary settlement between two peers. In other words, peering is free. A key difference between transit and peering is that transit provides full routes, whereas peering provides partial routes.

Suppose ISP-A (with 1,000 routes in its network) and ISP-B (with 1,500 routes in its network) are peering with each other. ISP-A provides ISP-B with all its routes (1,000 routes) via BGP. In return, ISP-B provides ISP-A with all its routes (1,500 routes) via BGP. Since ISP-A has learned all the routes of ISP-B, ISP-A will be able to send traffic to ISP-B and vice versa. In other words, peering is a relationship whereby two ISPs provide reciprocal access to each other's network.

THE INTERNET HIERARCHY

The Internet is generally composed of three layers: Tier 1 providers, Tier 2 providers, and End Users (enterprises, gaming companies, social networks, content providers, and so on).

Tier 1 providers: Tier 1 providers are at the top of the Internet hierarchy and they can maintain full routes or have global access just by peering with other Tier 1 providers. Generally Tier 1 providers have full mesh interconnections with each other as shown in the schematic diagram in Figure 1.

Note that the number of Tier 1 providers is extremely limited: approximately ten worldwide. As mentioned earlier, as of today 36,047 ISPs are connected, and together they constitute the Internet. This means only *0.03 percent* (equal to 10/36,047) of ISPs are Tier1 providers, and the rest are Tier 2 providers.

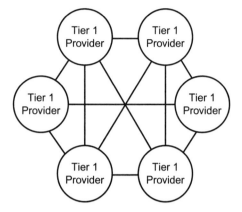

Figure 1. Tier 1 provider interconnection model.

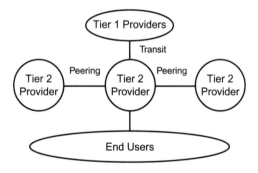

Figure 2. Tier 2 provider model.

Tier 2 providers: Tier 2 providers are ISPs who peer with other Tier 2 providers, but which still need to purchase transit service from Tier 1 providers (see Figure 2); 99.97 percent of ISPs fall under this category.

WHO PEERS? WHY PEER?

This section focuses on Tier 2 providers; they need peering on a much greater scale than Tier 1 providers. Tier 2 providers can obtain global reachability by simply purchasing full routes from Tier 1 providers and sending all their traffic to the Tier 1 providers. They need to peer for primarily three reasons: reduce transit costs, improve performance, and control routing.

Reduce transit costs: Tier 2 providers purchase transit service from Tier 1 providers by megabits per second. This means the more traffic they send to Tier 1 providers, the more they have to pay. On the other hand, peering is free. Considering the ever-increasing price pressure from competition in the market, what would you do if you were a Tier 2 provider to maintain a healthy profit level?

Probably, you would like to peer with as many Tier 2 providers as possible because peering is free. As Figure 3 shows, doing so would reduce the amount of traffic to Tier 1 providers and thus reduce transit costs. For instance, if you have 100 Gbps traffic in your network, your goal may be to send at least 20 percent (or 20 Gbps) of the traffic to peering and send 80 percent (or 80 Gbps) or less of traffic to Tier 1 providers. This would result in a significant cost reduction.

Improve performance: Tier 2 providers purchase full routes from Tier 1 providers. Full routes contain global *AS path* information. The AS path information is the information the route traffic must take based on the ASN. As Figure 4 shows, however, the AS path may not be optimal. If you are a Tier 2 provider and you want to send traffic to another Tier 2 provider, the destination Tier 2 provider may be multiple-AS paths away, resulting in higher latency. If your customer is a latency-sensitive customer such as an online gaming company, you may lose the customer if the latency of your network is high.

If you peer with a destination Tier 2 provider, the provider will become adjacent to your network (i.e., you will have a direct connection) and latency will be greatly improved.

Control routing: If you are a Tier 2 provider peering with another Tier 2 provider, you have two paths to the destination Tier 2 provider via transit and peering. If one of the paths is not performing well, you can choose the alternate path. In other words, if they do not peer, Tier 2 providers will increase costs, degrade performance, and lose control of routing.

Therefore, peering is very important for Tier 2 providers to maintain their business.

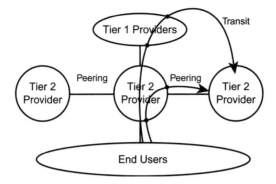

Figure 3. Traffic flow via transit versus peering.

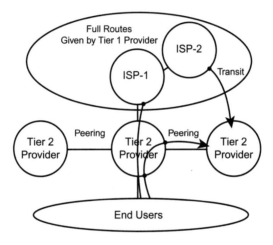

Figure 4. Traffic via transit not the shortest path.

A BRIEF HISTORY OF PEERING

Beginning with this section, we will look into peering from various per-spectives, starting with a business perspective. At the dawn of the Internet, all ISPs were competitors, but also friends. Most ISPs had an open peering policy where they agreed to peer with any other ISP with no prerequisites. At that time, one of the key sayings was "keep local traffic local." Before peering became common, local traffic from ISP-A to ISP-B was first sent to a Tier 1 provider from ISP-A and then the traffic came back to ISP-B via the Tier 1 provider. If ISP-A had been peering with ISP-B, the traffic would have been sent directly from ISP-A to ISP-B. Once ISPs realized the benefits of peering, they were virtually unanimous in supporting peering in general.

As the Internet grew, however, ISPs became more competitive. As a result, peering became a purely business decision. Some ISPs developed peering policies (or peering prerequisites) and started to peer selectively. Some ISPs *de-peered*, terminating existing peering relationships if they felt a particular peering (or peerings) did not make business sense. As of today, peering continues to be driven by business decisions, and ISPs carefully review and evaluate every single peering relationship periodically.

PEERING AS A BUSINESS DECISION

Every peering must make business sense. There are generally three types of peering from a business perspective: symmetrical peering, asymmetrical peering, and no customer peering.

Symmetrical peering: As Figure 5 shows, suppose ISP-A is in talks with ISP-B about creating a peering relationship. After estimating the traffic pattern after peering, ISP-A realizes that the traffic amount from ISP-B to ISP-A would be 1 Gbps, whereas the traffic from ISP-A to ISP-B would be 10 Mbps. This means ISP-B can reduce its transit costs significantly via the peering, but ISP-A cannot. In this case, the peering doesn't make business sense to ISP-A. If the traffic from ISP-A to ISP-B was 800 Mbps, it would make better business sense for ISP-A. In other words, peering makes sense if peering traffic is symmetrical between the two ISPs.

Asymmetrical peering: As Figure 6 shows, suppose ISP-A is in talks with ISP-B to have a peering relationship and ISP-A is a large content provider (e.g., a *content-heavy provider* such as an on-demand video provider) and ISP-B is a large regional operator (e.g., an *eyeball-heavy provider*, such as a local broadband provider).

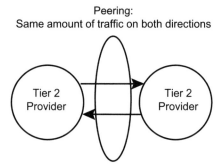

Figure 5. Symmetrical peering.

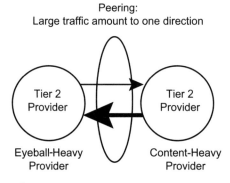

Figure 6. Asymmetrical peering.

Content-heavy providers would like to push traffic to eyeball-heavy providers (i.e., residential customers) while reducing transit costs. Eyeball-heavy providers would like to pull traffic from content-heavy providers for their residential customers while reducing transit costs.

Obviously the amount of traffic flowing from content-heavy providers to eyeball-heavy providers is always exponentially larger. Does it make sense to peer even though the traffic is not symmetrical? Of course it does. Both content-heavy providers and eyeball-heavy providers could reduce by a significant amount their transit costs by peering with each other.

At the time of this writing, news emerged of an eyeball-heavy provider (specifically, a cable operator) starting to charge a content-heavy provider for "peering" in order to protect its own cable television revenue. Peering is a dynamic process.

No customer peering: Suppose ISP-A is in talks with ISP-B to have a peering relationship. If ISP-A is a large ISP who serves North America nationwide and ISP-B is a small local ISP who serves a small city, ISP-B can be a customer of ISP-A. If ISP-A peers with ISP-B, ISP-A loses a revenue opportunity. In other words, ISPs won't peer with an ISP who is already their customer or potentially their customer.

BILATERAL VERSUS MULTILATERAL PEERING

This section looks at peering from the perspective of the peering negotiations involved, specifically looking at two kinds of peerings: bilateral and multilateral.

Bi-lateral peering: This type of peering requires peering negotiation between two ISPs. If ISP-A wishes to peer with ISP-B, ISP-A needs to talk with ISP-B. Both ISPs must agree to establish the peering relationship. If ISP-A wishes to peer with fifty ISPs, ISP-A needs to negotiate with fifty ISPs individually. From a technical perspective, bilateral peering means BGP settings on a border router need to be configured for every single peering relationship. Most ISPs prefer bilateral peering because they can peer selectively based on business decisions.

Multi-lateral peering: Multilateral peering requires no peering negotiation. If ISP-A agrees with multilateral peering, it means ISP-A agrees to peer with all ISPs. From a technical perspective, only one BGP setting is required to peer with all ISPs. Small ISPs or content providers who wish to peer with any ISP prefer multilateral peering.

PRIVATE VERSUS PUBLIC PEERING

This section looks at peering from the perspective of the physical connections involved.

Private peering: Private peering is used when two ISPs (and only two ISPs) peer with each other over a cable. They do not share the cable with other parties. In this case, it is each ISP's responsibility to find other ISPs to peer with.

Public peering: Public peering is used when an ISP peers with other ISPs on Internet Exchange Points (IXPs). IXPs use a *shared fabric* such as a layer 2 switch as a platform. Usually, the organization administering the IXP maintains a website providing information such as a list of participants, interface speed options, traffic statistics, hardware platform used, and so on.

Most ISPs prefer public peering at IXPs because they can peer easily and cost effectively.

Ease of peering: When one ISP wants to peer with another ISP privately, the ISP initiating the peering needs to spend a lot of time and effort to find answers to basic questions such as:

Where is the other ISP located? Is it in the same building or a different building?

What is the distance from us to them?

How do we connect?

Are there any local regulations governing running cables?

How much would it cost to build an interconnection?

At an IXP, these questions have been answered already. There is a participant list, so you know who is participating at the IXP. Since other participants have been already connected to the IXP, you care only about how to connect your router to the IXP, which, one way or another, is always achievable. Cost to participate in the IXP is easy to determine. Consequently, it is relatively easy to peer with other ISPs at any IXP.

Cost effectiveness: In private peering, an interconnection via cable is required for each peering. Each ISP will incur a cost to do so. For example, given that current Ethernet ports are either 1 Gbps or 10 Gbps, if an ISP needs to peer with four ISPs via private peering at speed of 2 Gbps, the ISP has to have four 10 GigE ports or eight GigE ports.

On the other hand, at an IXP, the ISP needs only one 10 GigE port to peer with four ISPs at a speed of 2 Gbps on the IXP, because

4×2 Gbps is less than 10 Gbps, so one 10 GigE connection to the IXP will suffice. At IXPs, peering is cost effective because an ISP can peer with multiple ISPs via a single router interface. In summary, Tier 2 providers need to peer on IXPs to maintain their business competitiveness.

A Brief History of IXPs

The first IXP was Commercial Internet Exchange (CIX, pronounced "kicks"), founded in 1991 by PSINet, AlterNet, and CERFNet. The hardware platform was a Cisco 7010 router located in Palo Alto, California, and managed by a not-for-profit association. CIX was an IXP pioneer and established the basic concept and business model.

Not long after, Metropolitan Area Ethernet (MAE, pronounced "may") and NSF-sponsored network access points (NAPs) were established by MCI and Sprint. They were managed by organizations for profit. Since then the concept and business model of IXPs has been well understood by the Internet community, and many IXPs have been established. Thanks to the efforts of many researchers and contributors in the Internet community, it is believed that there are somewhere between 300 and 350 IXPs worldwide [3]. However, the number of IXPs and the aggregated traffic on them is still growing.

Depending on countries or political boundaries, IXPs are operated by organizations for profit, non-profit organizations, or governments. Peering policy can be bi-lateral (can peer freely) or mandatory multi-lateral (must peer with all participants).

There are two types of IXPs: layer 2 and layer 3. A layer 2 IXP uses a shared network fabric like Ethernet switches as a platform for performing the exchange (see Figure 7). The IXP is not involved in participant routing. Participants determine who they peer with. They do not have to peer with everyone (bilateral peering).

A layer 3 IXP uses routers as the platform for performing the exchange (see Figure 8). The IXP is involved in participant routing. Participants have limited control over who to peer with, which could potentially create business issues. A layer 3 IXP's policy is that participants must peer with everyone (multilateral peering).

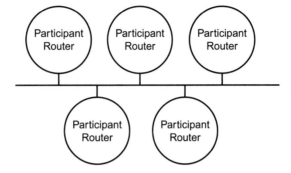

Figure 7. Layer 2 IXP architecture.

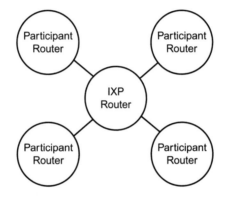

Figure 8. Layer 3 IXP architecture.

Today, most IXPs select layer 2 Ethernet switches as their platform for many reasons:

They are a proven platform for many years
They support high speeds: 1 G/10 G/100 G
The hardware is relatively cheap
They are easy to operate (no BGP operation)
They offer optional private peering via VLAN

A layer 2 switch platform is a bilateral peering environment. However, some IXPs provide a route server to offer multilateral peering capability in addition to the default bilateral peering environment.

IXPS AND CYBER SECURITY

Because there are over three hundred IXPs in the world and they carry a tremendous amount of important Internet traffic, it is important that IXPs are well protected against cyber attacks. Because of this, the IP address range of the public peering network must not be announced by any participants of IXPs to ensure the IP address range is not reachable from anywhere. Periodic vulnerability checks must be performed on IXP platforms to ensure all software is vulnerability free. IXP platforms must be Communications Assistance for Law Enforcement (CALEA) compliant to facilitate investigation processes by law enforcement.

TODAY'S CHALLENGE

While layer 2 switches are the platform of choice for most IXPs, the platform is not perfect and has several challenging issues, among which are the following:

Ever-increasing traffic: The aggregated traffic on IXPs has grown significantly in the past ten years and it is still growing. In the past ten years, most popular ports transitioned from 100 MB, 1 GB to 10 GB. At large IXPs, it has been truly a challenge to accommodate hundreds of 10 GB port participants. Large IXPs hope that Ethernet switch manufactures continue to increase switch capacity and port density to meet ever-increasing market demands.

Not fully secure: As all participants are in the same broadcast domain, peering traffic can be affected by other participants' traffic. Using a private connection (VLAN) per peer is not practical because the number of VLANs is limited to 4,096. As an example of the effect of this limitation, if 120 participants are peering in a meshed configuration, 7,140 VLANs are necessary ($= 120 \times 120 - 1)/2$). It is also difficult to manage such a large number of VLANs.

No MTU optimization: MTU (maximum transmission unit) is the largest layer 2 protocol data unit supported. In the case of Ethernet, the MTU is 1,500 bytes. The 1,500 bytes of MTU were designed in the 1980s when Ethernet's speed was 10 Mbps. Today, Ethernet's speed is 10 Gbps or faster. It would be more efficient to adopt a larger MTU size (*jumbo frames*) when using these higher speeds. However, it is difficult to choose a larger MTU that is agreeable to all participants because every participant's router or network has a different optimal MTU size. Therefore, most IXPs still use the default MTU size of 1,500 bytes.

No traffic statistics per peer: In order to collect statistics, we rely on SNMP/MIB (Simple Network Management Protocol/Management Information Base). Currently MIB is available per physical switch port. However, this is not convenient because only traffic statistics per physical port can be collected. Since multiple peering sessions will pass through one physical port, it would be beneficial if statistics can be collected per peering session.

While today there is a technology called sFlow to collect statistics per peering session, this is a sampling-based method and thus the statistics are not 100 percent accurate.

FUTURE IXP ARCHITECTURE

Layer 2 switch platforms are likely to be replaced with newer technologies. One of the possibilities is an MPLS-based IXP. I presented the concept at a NANOG meeting in 2002. The MPLS-based IXP has the potential to overcome several drawbacks of layer 2 switches. Notably, it offers secure tunnel-per-peer, optimal MTU-per-peer, SNMP MIB-per-tunnel, and traffic engineering via RSVP.

Conclusion

This chapter provides a basic degree of information about IXPs and their role in cyberspace, as summarized below:

The Internet is a network of multiple autonomous networks, and is composed of over 35,000 ISPs.

BGP connects ISPs.

ISPs are connected either by transit or by peering relationships.

Peering is a business decision.

Many peering types exist (bilateral, multilateral, private, public, and so on).

Tier 2 providers need to peer on IXPs to reduce costs, increase performance, and control routes. Without IXPs, it is difficult for Tier 2 providers to maintain their business and competitiveness.

There are over 300 IXPs worldwide. The number of IXPs and their aggregated traffic is still growing.

Depending on countries or political boundaries, IXPs are operated by organizations for profit, non-profit organizations, or government.

Most IXPs are based on a layer 2 switch platform. However, there are some challenges as a result of this situation.

In the future, IXP may be based on newer hardware architecture, and MPLS-IXP could be one of the options.

ORGANIZATIONS SUPPORTING THE INTERNET

Due to the multi-autonomous nature of the network, the Internet is difficult to maintain. To help with this, there are numerous organizations supporting the Internet. In other words, the Internet exists because of efforts or contributions by many people. The following is a select list—in no particular order—of organizations that support the Internet in various ways, but it is by no means comprehensive.

Internet Corporation for Assigned Names and Numbers (ICANN): http://www.icann.org/

Internet Assigned Numbers Authority (IANA): http://www.iana.org/

Internet Society (ISOC): http://www.internetsociety.org/

Internet Engineering Task Force (IETF): http://www.ietf.org/

International Telecommunication Unit—Telecommunication Standardization Sector (ITU-T): http://www.itu.int/ITU-T/

American Registry for Internet Numbers (ARIN): https://www.arin.net/

Réseaux IP Européens Network Coordination Centre (RIPE NCC): http://www.ripe.net/

Asian-Pacific Network Information Center (APNIC): http://www.apnic.net/

Latin American and Caribbean Network Information Center (LAC-
NIC): http://lacnic.net/en/index.html
Association Française pour le Nommage Internet en Coopération
(AFNIC): http://www.afnic.fr/en/
Internet Software Consortium (ISC): http://www.isc.org/
Asia Pacific Regional Internet Conference on Operational Technologies
(APRICOT): http://www.apricot.net/
North American Network Operators' Group (NANOG): http://www
.nanog.org/
African Network Operators' Group (AFNOG): http://www.afnog.org/
French Network Operators' Group (FRNOG): http://www.frnog.org/
New Zealand Network Operators' Group (NZNOG): http://www
.nznog.org/
Japan Network Operators' Group (JANOG): http://www.janog.gr.jp/en/

REFERENCES

[1] CIDR Report page. [Online]. Available: http://www.cidr-report.org
[2] RFC 1771 A Border Gateway Protocol 4. [Online]. Available: http://www
.ietf.org/rfc/rfc1771.txt
[3] Peering DB. [Online]. Available: http://www.peeringdb.com

Part II: Operations

Tor: Uses and Limitations of Online Anonymity

Andrew Lewman

Tor is a network of virtual tunnels that allows people and groups to improve their privacy and security on the Internet. It also enables software developers to create new communication tools with built-in privacy features. Tor provides the foundation for a range of applications that allow organizations and individuals to share information over public networks without compromising their privacy.

Individuals use Tor to keep websites from tracking them and their family members, or to connect to news sites, instant messaging services, or the like when these are blocked by their local Internet providers. Tor's hidden services let users publish websites and other services without needing to reveal the location of the site. Individuals also use Tor for socially sensitive communication: chat rooms and web forums for rape and abuse survivors, or people with serious illnesses.

Journalists use Tor to communicate more safely with whistle-blowers and dissidents. Nongovernmental organizations (NGOs) use Tor to allow their workers to connect to their home website while they're in a foreign country, without notifying everybody nearby that they're working with that organization.

Groups such as Indymedia recommend Tor for safeguarding their members' online privacy and security. Activist groups like the Electronic Frontier Foundation (EFF) recommend Tor as a mechanism for maintaining civil liberties online. Corporations use Tor as a safe way to conduct competitive analysis, and to protect sensitive procurement patterns from eavesdroppers. They also use it to replace traditional VPNs, which reveal the exact amount and timing of communication. Which locations have employees working late? Which locations have employees consulting job-hunting websites? Which research divisions are communicating with the company's patent lawyers?

A branch of the U.S. Navy uses Tor for open source intelligence gathering, and one of its teams recently used Tor while deployed in the Middle East. Law enforcement uses Tor for visiting or surveilling web sites without leaving government IP addresses in their web logs, and for security during sting operations.

The variety of people who use Tor is actually part of what makes it so secure. Tor hides you among the other users on the network, so the more populous and diverse the user base for Tor is, the more your anonymity will be protected.

Why We Need Tor

Using Tor protects you against a common form of Internet surveillance known as *traffic analysis*. Traffic analysis can be used to infer who is talking to whom over a public network. If others know the source and destination of your Internet traffic, they can track your behavior and interests. This can impact your checkbook if, for example, an e-commerce site uses price discrimination based on your country or institution of origin. It can even threaten your job and physical safety by revealing who and where you are. For example, if you're travelling abroad and you connect to your employer's computers to check or send mail, you can inadvertently reveal your national origin and professional affiliation to anyone observing the network, even if the connection is encrypted.

How does traffic analysis work? Internet data packets have two parts: a data payload and a header used for routing. The data payload is whatever is being sent, whether that's an e-mail message, a web page, or an audio file. Even if you encrypt the data payload of your communications, traffic analysis still reveals a great deal about what you're doing and, possibly, what you're saying. That's because it focuses on the header, which discloses source, destination, size, timing, and so on.

A basic problem for the privacy minded is that the recipient of your communications can see, by looking at headers, that you sent it. So can authorized intermediaries like Internet service providers, and sometimes unauthorized intermediaries as well. A very simple form of traffic analysis might involve sitting somewhere between sender and recipient on the network, looking at headers.

But there are also more powerful kinds of traffic analysis. Some attackers spy on multiple parts of the Internet and use sophisticated statistical techniques to track the communications patterns of many different organizations and individuals. Encryption does not help against these attackers, since it hides only the content of Internet traffic, not the headers.

Tor helps to reduce the risks of both simple and sophisticated traffic analysis by distributing your transactions over several places on the Internet, so no single point can link you to your destination. The idea is similar to using a twisty, hard-to-follow route in order to throw off somebody who is tailing you—and then periodically erasing your footprints. Instead of taking a direct route from source to destination, data packets on the Tor network take a random pathway through several relays that cover your tracks so no observer at any single point can tell where the data came from or where it's going.

To create a private network pathway with Tor, the user's software or client incrementally builds a circuit of encrypted connections through relays on the network. The circuit is extended one hop at a time, and each relay along the way knows only which relay gave it data and which relay it is giving data to. No individual relay ever knows the complete path that a data packet has taken. The client negotiates a separate set of encryption keys for each hop along the circuit to ensure that each hop can't trace these connections as they pass through.

Once a circuit has been established, many kinds of data can be exchanged, and several different sorts of software applications can be deployed over the Tor network. Because each relay sees no more than one hop in the circuit, neither an eavesdropper nor a compromised relay can use traffic analysis to link the connection's source and destination. Tor works only for TCP streams and can be used by any application with SOCKS support.

For efficiency, the Tor software uses the same circuit for connections that happen within the same ten minutes or so. Later requests are given a new circuit to keep people from linking your earlier actions to the new ones.

HIDDEN SERVICES

Tor also makes it possible for users to hide their locations while offering various kinds of services, such as web publishing or an instant messaging server. Using Tor "rendezvous points," other Tor users can connect to these hidden services, each without knowing the other's network identity. This hidden service functionality could allow Tor users to set up a website where people publish material without worrying about censorship. Nobody would be able to determine who was offering the site, and nobody who offered the site would know who was posting to it.

STAYING ANONYMOUS

Tor can't solve all anonymity problems. It focuses only on protecting the transport of data. You need to use protocol-specific support software if you don't want the sites you visit to see your identifying information. For example, you can use web proxies such as Privoxy while web browsing to block cookies and withhold information about your browser type.

Also, to protect your anonymity, be smart. Don't provide your name or other revealing information in web forms. Be aware that, like all anonymizing networks that are fast enough for web browsing, Tor does not provide protection against end-to-end timing attacks: If your attacker can watch the traffic coming out of your computer, and also the traffic arriving at your chosen destination, he can use statistical analysis to discover that they are part of the same circuit.

THE FUTURE OF TOR

Providing a usable anonymizing network on the Internet today is an ongoing challenge. We want software that meets users' needs. We also want to keep the network up and running in a way that handles as many users as possible. Security and usability don't have to be at odds: As Tor's usability increases, it will attract more users, which will increase the possible sources and destinations of each communication, thus increasing security for everyone.

Ongoing trends in law, policy, and technology threaten anonymity as never before, undermining our ability to speak and read freely online. These trends also undermine national security and critical infrastructure by making communication among individuals, organizations, corporations, and governments more vulnerable to analysis. Each new user and relay provides additional diversity, enhancing Tor's ability to put control over your security and privacy back into your hands.

Typical Tor Usage

Normal people use Tor.

People use Tor to protect their privacy from unscrupulous marketers and identity thieves. Internet service providers (ISPs) sell your Internet browsing records to marketers or anyone else willing to pay for them. ISPs typically say that they anonymize the data by not providing personally identifiable information, but this has been proven to be incorrect. A full record of every site you visit, the text of every search you perform, and potentially your user ID and even your password information can still be part of this data. In addition to your ISP, the websites (and search engines) you visit have their own logs, containing the same or more information.

By using Tor, people protect their communications from irresponsible corporations. All over the Internet, Tor is being recommended to people concerned about their privacy in the face of increasing breaches and betrayals of private data. The problems range from lost backup tapes to data being given away to researchers—your data is often not well protected by those you are supposed to trust to keep it safe.

Parents protect their children online. You've told your kids they shouldn't share personally identifying information online, but they may be sharing their location simply by not concealing their IP address. Increasingly, IP addresses can be literally mapped to a city or even street location, and can reveal other information about how you are connecting to the Internet. In the United States, the government is pushing to make this mapping increasingly precise.

Individuals research sensitive topics. There's a wealth of information available online. But perhaps in your country, access to information on AIDS, birth control, Tibetan culture, or world religions is behind a national firewall.

MILITARIES USE TOR

Tor's anonymity is very valuable to the military in many ways. It is not difficult for insurgents to monitor Internet traffic and discover all the hotels and other locations from which people are connecting to known military servers. Military field agents deployed away from home use Tor to mask the sites they are visiting, protecting military interests and operations, as well as protecting themselves from physical harm.

When the Internet was designed by DARPA, its primary purpose was to be able to facilitate distributed, robust communications in case of local strikes. However, some functions must be centralized, such as command and control sites. It's the nature of Internet protocols to reveal the geographic location of any server that is reachable online. Tor's hidden services capacity allows military command and control to be physically secure from discovery and takedown.

Military personnel need to use electronic resources run and monitored by insurgents. They do not want the web server logs on an insurgent website to record a military address, thereby revealing the surveillance.

JOURNALISTS AND THEIR AUDIENCE USE TOR

Reporters without Borders tracks Internet prisoners of conscience and jailed or harmed journalists all over the world. They advise journalists, sources, bloggers, and dissidents to use Tor to ensure their privacy and safety. The U.S. International Broadcasting Bureau (Voice of America/Radio Free Europe/Radio Free Asia) supports Tor development to help Internet users in countries without safe access to free media. Tor preserves the ability of persons behind national firewalls or under the surveillance of repressive regimes to obtain a global perspective on controversial topics including democracy, economics, and religion.

Citizen journalists in China use Tor to write about local events to encourage social change and political reform.

Citizens and journalists in Internet black holes use Tor to research state propaganda and opposing viewpoints, to file stories with non-state controlled media, and to avoid risking the personal consequences of intellectual curiosity.

LAW ENFORCEMENT OFFICERS USE TOR

For law enforcement involved with online surveillance, sting operations, or protecting anonymous sources, Tor adds the safety level needed for success in fighting crime.

Tor allows officials to surf questionable web sites and services without leaving tell-tale tracks. If the system administrator of an illegal gambling site, for example, were to see multiple connections from government or law enforcement IP addresses in usage logs, investigations could be hampered.

Similarly, anonymity allows law officers to engage in online undercover operations. Regardless of how good an undercover officer's "street cred" may be, if the communications include IP ranges from police addresses, the cover is blown.

Truly anonymous tip lines are also possible. While online anonymous tip lines are popular, without anonymity software, they are far less useful than they could be. Sophisticated sources understand that although a name or email address is not attached to information, server logs can identify them very quickly. As a result, tip line web sites that do not encourage anonymity are limiting the sources of their tips.

ACTIVISTS AND WHISTLE-BLOWERS USE TOR

Human rights activists use Tor to anonymously report abuses from danger zones. Internationally, labor rights workers use Tor and other forms of online and offline anonymity to organize workers in accordance with the Universal Declaration of Human Rights. Although they are within the law, it does not mean they are safe. Tor provides the ability to avoid persecution while still raising a voice.

When groups such as the American Friends Service Committee and environmental groups are increasingly falling under surveillance in the United States under laws meant to protect against terrorism, many peaceful agents of change rely on Tor for basic privacy during legitimate activities.

Human Rights Watch recommends Tor in their report *"Race to the Bottom": Corporate Complicity in Chinese Internet Censorship.* The study's co-author interviewed Roger Dingledine, Tor project leader, on Tor use. They cover Tor in the section on how to breach the "Great Firewall of China," and recommend that human rights workers throughout the globe use Tor for "secure browsing and communications."

Tor has consulted with and volunteered help to Amnesty International's recent corporate responsibility campaign. Global Voices recommends Tor, especially for anonymous blogging, throughout their web site.

In the United States, the Supreme Court recently stripped legal protections from government whistle-blowers. But whistle-blowers working for governmental transparency or corporate accountability can use Tor to seek justice without personal repercussions.

A contact of mine who works with a public health nonprofit in Africa reports that his nonprofit must budget 10 percent to cover various sorts of corruption, mostly bribes and such. When that percentage rises steeply, not only can they not afford the money, but they can not afford to complain—this is the point at which open objection can become dangerous. So, to continue their work, his nonprofit has been working to use Tor to safely expose government corruption.

At a recent conference, a Tor staffer ran into a woman who came from a "company town" in the eastern United States. She was attempting to blog anonymously to rally local residents to urge reform in the company that dominated the town's economic and government affairs. She is fully cognizant that the kind of organizing she was doing could lead to harm or fatal "accidents." In East Asia, some labor organizers use anonymity to reveal information regarding sweatshops that produce goods for western countries and to organize local labor.

Tor can help activists avoid government or corporate censorship that hinders organization. In one such case, a Canadian ISP blocked access to a union website used by its own employees to help organize a strike.

Does being in the public spotlight shut you off from having a private life, forever, online? A rural lawyer in a New England state keeps an anonymous blog because, with the diverse clientele at his prestigious law firm, his political beliefs are bound to offend someone. Yet he doesn't want to remain silent on issues he cares about. Tor helps him feel secure about expressing his opinion without consequences to his public role.

People living in poverty often don't participate fully in civil society—not out of ignorance or apathy, but out of fear. If something you write were to get back to your boss, would you lose your job? If your social worker read about your opinion of the system, would she treat you differently? Anonymity gives a voice to the voiceless. Although it's often said that the poor do not use online access for civic engagement, thereby failing to act in their self-interest, it is my hypothesis (based on personal conversations and anecdotal information) that it is precisely the "permanent record" left online that keeps many of the poor from speaking out on the Internet. The hope is that Tor will show people how to engage more safely online, and then at the end of the year evaluate how online and offline civic engagement has changed, and how the population sees this continuing into the future.

BUSINESS EXECUTIVES USE TOR

Tor is a versatile tool for business as part of safety protocol for security clearinghouses, monitoring competition, and maintaining accountability.

Say a financial institution participates in a security clearinghouse of information on Internet attacks. Such a repository requires members to report breaches to a central group that correlates attacks to detect coordinated patterns and sends out alerts. But if a specific bank in St. Louis is breached, they don't want an attacker watching the incoming traffic to such a repository to be able to track where information is coming from. Even with every packet encrypted, the IP address would betray the location of a compromised system. Tor allows such repositories of sensitive information to resist becoming compromised.

If you try to check out a competitor's pricing, you may find no information or misleading information on their website. This is because their web server may be keyed to detect connections from competitors, and block access by or spread disinformation to your staff. Tor allows a business to view their sector as the general public would view it.

Tor helps to keep strategies confidential. An investment bank, for example, might not want industry snoopers to be able to track what web sites their analysts are watching. The strategic importance of traffic patterns, and the vulnerability of the surveillance of such data, is starting to be more widely recognized in several areas of the business world.

In an age when irresponsible and unreported corporate activity has undermined multi-billion dollar businesses, an executive exercising true stewardship wants the whole staff to feel free to disclose internal malfeasance. Tor facilitates internal accountability before it turns into whistle-blowing.

Advanced Concepts

HIDDEN SERVICES

Tor makes it possible for users to hide their locations while offering various kinds of services, such as web publishing or an instant messaging server. Using Tor rendezvous points, other Tor users can connect to these hidden services, each without knowing the other's network identity.

A hidden service needs to advertise its existence in the Tor network before clients can contact it. Therefore, the service randomly picks some relays, builds circuits to them, and asks them to act as introduction points by telling them its public key. By using a full Tor circuit, it's hard for anyone to associate an introduction point with the hidden server's IP address. Although the introduction points and others are told the hidden service's identity (public key), we don't want them to learn about the hidden server's location (IP address).

The hidden service assembles a hidden service descriptor, containing its public key and a summary of each introduction point, and signs this descriptor with its private key. It uploads that descriptor to a distributed hash table. The descriptor will be found by clients requesting *XYZ*.onion where *XYZ* is a sixteen character name that can be uniquely derived from the service's public key. After this step, the hidden service is set up.

Although it might seem impractical to use an automatically-generated service name, it serves an important goal: Everyone—including the introduction points, the distributed hash table directory, and of course the clients—can verify that they are talking to the right hidden service. (Just remember Zooko's conjecture that of the properties *decentralized, secure,* and *human-meaningful,* you can achieve at most two. Perhaps one day somebody will implement a "Petname" design for hidden service names.)

In general, the complete connection between client and hidden service consists of six relays: three of them were picked by the client with the third being the rendezvous point and the other three were picked by the hidden service.

Some examples of active hidden services are http://56apzofkmsmgb3yr .onion/, which provides access to http://archive.torproject.org, and http:// 3g2upl4pq6kufc4m.onion/, which is a privacy-preserving way to access the DuckDuckGo Search engine (https://duckduckgo.com/).

RESEARCH

Many people around the world are doing research on how to improve the Tor design, what's going on in the Tor network, and more generally on attacks and defenses for anonymous communication systems. One resource to use for starting research into anonymous communication systems is Freehaven's Anonbib (http://freehaven.net/anonbib/).

We've been collecting data to learn more about the Tor network: how many relays and clients there are in the network, what capabilities they have, how fast the network is, how many clients are connecting via bridges, what traffic exits the network, and so on. We are also developing tools to process these huge data archives and come up with useful statistics. For example, we provide a tool called Ernie that can import relay descriptors into a local database to perform analyses. Our metrics website, https://met rics.torproject.org, is one such respository.

These days we review too many conference paper submissions that make bad assumptions and end up solving the wrong problem. Since the Tor protocol and the Tor network are both moving targets, measuring things without understanding what's going on behind the scenes is going to result in bad conclusions. In particular, different groups often unwittingly run a variety of experiments in parallel, and at the same time we're constantly modifying the design to try new approaches.

We're building a repository of tools that can be used to measure, analyze, or perform attacks on Tor. Many research groups need to do similar measurements (for example, change the Tor design in some way and then see if latency improves), and we hope to help everybody standardize on a few tools and then make them really good. Also, while people have published about some neat Tor attacks, it's hard to track down a copy of the code they used.

Most researchers find it easy and fun to come up with novel attacks on anonymity systems. Lately we've seen this result in improved congestion attacks, attacks based on remotely measuring latency or throughput, and so on. Knowing how things can go wrong is important, and we recognize that the incentives in academia aren't aligned with spending energy on designing defenses, but it sure would be great to get more attention on how to address the attacks.

Conclusion

Tor is used the world over to allow people to protect their anonymity and privacy online. It's an active research network, testing ground for new anonymity and privacy technologies and designs, and is relied upon by millions of people in nearly every country in the world.

It's used to protect sensitive communications between individuals, allow access to an uncensored Internet from within oppressive regimes, and allow others to simply use unsafe networks to access secure destinations. This diversity of users and uses is what gives Tor its anonymizing properties and creates a larger crowd in which an individual can hide.

Join us in improving, researching, and using Tor!

Authoritative Data Sources:
Cyber Security Intelligence Perspectives

Kuan-Tsae Huang and Hwai-Jan Wu

Many reports indicate that the United States currently faces a multifaceted, technologically based vulnerability. Our information systems are being exploited on an unprecedented scale by state and non-state actors, resulting in a dangerous combination of known and unknown vulnerabilities, strong adversary capabilities, and weak situational awareness. Many cyber security issues arose due to the lack of systematic management of authoritative data sources (ADS).

This chapter presents the concepts and applications of ADS, authoritative data elements, and trusted data sources to enable rapid business processes. It also discusses trusted intelligent decision making in network-centric business environments. We describe a methodology to standardize the ADS processes by which we manage, use, and secure ADS and apply various applications to provide cyber security.

Actionable Intelligence

Actionable intelligence is derived from actionable information—information that can be acted on—collected from trusted and authoritative sources. It is something that can lead to an action that can trigger a chain of other actions and reactions. Usually, actionable information is a relatively small piece (or pieces) of non-obvious details that can form an initial basis for hypothesis building. Actionable information is the cornerstone of decision support. The process of deriving actionable intelligence from actionable information is a matter of analyzing and processing that information to arrive at certain decision support hypotheses that can lead to action. Arriving at actionable intelligence could involve using various matrices and methodologies to process a number of bits of information that intelligence agencies might have received from a variety of sources.

Multiple data sources containing potentially inaccurate and outdated copies of data can be confusing, and can be time-consuming and expensive to maintain. Developing a registry that identifies the single trusted—or authoritative—data source will save time as well as the costs associated with maintaining and verifying the accuracy of multiple sources.

The data on its own may or may not be intelligible until transformed. This need to express data as intelligible information necessitates data abstraction. Data abstraction is accomplished by separating raw data into understandable and cohesive pieces, then transforming those independent pieces of raw data into intelligible information. As the single resource to obtain reliable and trustworthy information, ADS must have associated data abstraction, which is critical to coherent, consistent, and enterprise-wide data management. According to the Department of Defense (DoD), an ADS is "a recognized or official data production source with a designated mission statement or source/product to publish reliable and accurate data for subsequent use. An ADS may be the functional combination of multiple or separate data sources" [9].

CYBER SECURITY THREATS

After the Christmas Day bombing attempt on Flight 253, an analysis discovered that a mix of human and systemic failures with handling ADS contributed to this potentially catastrophic breach of security. The underwear bomber, Abdulmutallab, was listed in the Terrorist Identities Datamart Environment (TIDE) owned by the National Counterterrorism Center. That database stores information about one million individuals who may be a threat to the United States. By examining the TIDE database, the FBI Terror Screening Unit identifies those who are the real threat and places them in their Terrorism Screening Database, which contains records on 400,000 individuals. The FBI passes these names to the Transportation Security Administration (TSA), which maintains its own no-fly list of about 14,000 people.

An ADS, also called a trusted data source, is one where you know that not only can you trust the source, but you can also trust that you are looking at the exact same data others are. Problems arise from how the data is shared or, in the case of the underwear bomber, not shared. The common practice of copying data, sending it to others, and calling it "sharing" is really where this case went wrong: The data you copy is trusted data, but no authority other than you has it. Because it has been copied, the burden is on the recipient to ensure it has not been tampered with and that it is up-to-date.

Incorrect management of ADS is another cause of problems. There is no way for the TSA to combine its information about an individual who purchased a ticket in cash and has no luggage with the fact that his father reported him to the U.S. Embassy in Nigeria. The following is a list of areas for improving the methodology for managing the ADS infrastructure:

Allow TSA analysts with the proper credentials access to the information. Understand that the FBI's Terrorist Screening Database (TSDB) and the TSA's no-fly list are no longer ADS unless they share a real-time connection.

Combining and correlating data must be done by an analyst using appropriate tools.

Support real-time bi-directional analysis among data sources— connections are permitted, but ensure appropriate constraints, or limits, are placed on the data based on the user's credentials and on security policies.

It is overwhelming for most IT security staff to deal with multiple new vulnerability alerts each day. We need a solution allowing for more efficient management of this information by automatically consolidating in real time threat intelligence data from authoritative sources such as the National Vulnerability Database, commercial threat feeds, email advisories, and multiple zero-day and early warning services.

This chapter shows how to use the ADS infrastructure to define policy to clarify and specifically dictate from the beginning how data will be collected, made available, and managed so the communities needing the data can make necessary business decisions. It also shows the methodology to manage it. Policies should be discussed and agreed to by all parties following the ground rules for data rights management. Recognizing how important it is that ADS be provided through a trusted source is essential to protect data privacy and safety, and to ensure that the user community has the best available and most current information. The "Methodology for Managing Authoritative Data Sources" section describes the methodology to manage ADS. The "Formalizing Authoritative Data Sources" section organizes the concept in a formal paradigm with the potential for building a theoretical foundation in the future. The "Applications of Authoritative Data Sources" section refers to the National Spatial Data Infrastructure and the National Identity Ecosystem, which require leveraging the ADS Framework to manage various ADS and share identity and other attributes within the shared enterprise infrastructure. But securing complex heterogeneous environments is no easy task. Our work reported here is preliminary.

Methodology for Managing Authoritative Data Sources

An ADS has to follow a managed process to ensure data integrity in an application. When data are created or edited, data integrity is preserved by providing a single object through which updates can occur. When data are imported, which may be from multiple sources, there is a strict vetting process to ensure the integrity of the imported data. The process of managing the updates of records, from the systems designer's perspective, results in authoritative data. The ADS must provide accurate metadata that describes the lineage, quality, and currency for data elements.

There are four steps to achieve data integrity, accuracy, and quality for ADS in cyberspace:

1. Develop a comprehensive ADS Framework.
2. Build and implement an interoperable data control and configuration infrastructure aligned with the ADS Framework.
3. Enhance confidence and willingness to participate in the ADS infrastructure.
4. Ensure the life-cycle and long-term success of the ADS infrastructure.

The first two steps focus on designing and building the necessary governance and infrastructure to deliver online services securely. The third step targets the necessary privacy and security protections and the education and awareness required to encourage adoption. The fourth step establishes the structure and priorities to promote continued development and improvement of an ADS trusted system over time.

DEVELOP A COMPREHENSIVE ADS FRAMEWORK

In large enterprises, many different departments produce data that are geared toward their specific requirements. However, when data need to be shared across the enterprise, there are considerations beyond the specific needs of the organization. These include issues of trust, security, policy, understandability, quality, and so on. Data governance is the means to address these issues. The ADS Framework must form a community of interest (COI) to enable policy development and create robust practices for data governance and data assurance across communities. It guides the development of individual trusted data sources and addresses the barriers to make more effective and efficient use of data assets in achieving its operational goals.

The ADS Framework will include establishing comprehensive standards and committees based on defined risk models to manage the availability, usability, integrity, and security of enterprise data assets. Key elements of the ADS Framework are defining the rights and responsibilities of the various participants in the ADS, liability issues on privacy, compliance, and security, and establishing an enforcement mechanism in case participants do not carry out these responsibilities.

BUILD AND IMPLEMENT INTEROPERABLE DATA CONTROL AND CONFIGURATION
INFRASTRUCTURE ALIGNED WITH THE ADS FRAMEWORK

This step focuses on identifying authoritative data elements and their associated sources through a governance authority organized by COIs. When the same or similar data is stored in multiple sources, the possibility exists for some of the sources to contain inaccurate or outdated data, leading to potentially inaccurate decisions. It requires an infrastructure to support the interactions between data asset participants. One approach is to implement a registry sponsored by the COI to ensure that the most accurate and current data is managed, visible, trusted, and responsive to consumers. It will save time as well as the costs associated with maintaining and verifying the accuracy of multiple sources. However, this step requires promoting and incentivizing swift implementation of all levels of interoperability and widespread use among participants in the ADS infrastructure by all communities. The ADS infrastructure requires verification, validation, and certification management processes.

ENHANCE CONFIDENCE AND WILLINGNESS TO PARTICIPATE IN THE ADS SYSTEM

The ADS infrastructure must provide strong privacy and security protections to data assets in addition to creating clear rules and guidelines concerning the circumstances under which a service or data provider may share information and the kinds of information they may share. The confidence and willingness of user communities can be built by leveraging successes at the foundational data level, where data is produced and consumed, through the information level. At the information level, data is merged across applications and communities to enhance and expand the information needed for senior level decision makers to take and command actions. The approach is to promote confidence via mechanisms that address privacy protection, data integrity, and data confidentiality associated with ADS infrastructure. It will also address awareness of and education about both the risks associated with poor identification and authentication approaches and the ways in which the infrastructure mitigates those risks. These activities must leverage existing programs and engagement efforts and begin as soon as possible. They must also evolve as the ADS infrastructure matures to ensure that control and configuration processes are in alignment with the current environment.

This step involves exposing data to make data visible and accessible. Continued progress and innovation in data collection and the creation of an extended, trusted ADS infrastructure rely on significant community participation in organizational and community-wide standards development. Today's decision making environment is driven by a secure ADS infrastructure that can automatically respond to data requests to make data accessible and responsive to consumers' needs. The data services are developed based on a common semantic schema and in accordance with the data framework and data service layer specifications. Data warehouses can ingest the data (via the data services) for analysis, and security ensures the data is trusted and not compromised. To be successful, the organization must focus on technologies and research and development that have the potential to shift the security, reliability, resilience, and trustworthiness paradigm to benefit those who are depending on the data assets.

Formalizing Authoritative Data Sources

There is a difference between organizational authority and authoritative data. These concepts and terms need to be more fully described. An organization's declaration that a source is authoritative won't make it so. The implementation of the ADS relies on the source evolving through stewardship practices. Authority depends in part on who controls the data sources. How is the authority defined and what does the authority imply in an organization or application? How does being an ADS affect the database? How will this position affect existing and future partnerships? How will the review of data sets and their stewards affect the line of business of an organization? As with data management, we see a need to describe the concept of ADS in a formal paradigm with the potential for building a theoretical foundation in the future through logic and set theory. Note that this is strongly influenced by the DoD and Army experiences and is in an early stage of research and development. The idea described here is to establish a suitable form of communication, interpretation, and processing either by human or automated systems.

Under DoD policy, data is an essential enabler of network-centric warfare (NCW) and shall be made visible, accessible, and understandable to any potential users within the DoD as early as possible in its lifecycle to support mission objectives [1]. As part of the DoD's integrated enterprise-wide UID strategy for the development, management, and use of unique identifiers and their associated ADS, DoD Directive 8320.03 "Unique Identification (UID) Standards for a Net-Centric Department of Defense" defined the Authoritative Data Source as:

> a recognized or official data production source with a designated mission statement or source/product to publish reliable and accurate data for subsequent use by customers. An authoritative data source may be the functional combination of multiple, separate data sources.

This definition is the cornerstone for the Army's ADS Registry concept. However, it has some ambiguities that need to be defined for better understanding. We recognized that this ADS definition contains five attributes:

data production sources
designation statement for sources
publishing
reliable and accurate data for subsequent use by customers
functional combination of multiple, separate data sources

If we can formalize the definitions of these five attributes, then we can have a foundation to build an ADS.

Data Production Sources

A data source is a recognized or official collection of data that meets the specified requirements of a data consumer. Each data source has an information manager to manage the source database as well as handle the incoming requests.

A data source can be further classified into two types:

Originating data source: This is a data source that has a production process involved in generating the data from the real world. Examples include stock data, workforce tracking data—when and where a unit is in real time—and credit transaction data.

Intermediate data source: This is a data source that has a production process involved and is derived from originating/intermediate data sources for reporting purposes. Examples include data warehousing and business intelligence, which aggregates row data from original data sources.

Designation Statement for Sources

A designated mission statement is a formal, short, written statement of the purpose of a company or organization. The mission statement should guide the actions of the organization, spell out its overall goals, provide a sense of direction, and guide decision making. It provides the framework or context within which the company's strategies are formulated [6]. The mission statement of the DoD is providing the military forces needed to deter war and protect the security of the United States [7].

The designation statement for sources is a designated purpose for using data. One example is the BEACH data that was used for calculating Australian health welfare until 2010 [11]. In this case, BEACH data sources were used for the health welfare calculation purposes and during a designated period of time specified by the health department.

Publishing

A community of interest (COI) is defined as "a collaborative group of users who must exchange information in pursuit of their shared goals, interests, missions, or business processes and who therefore must have a shared vocabulary for the information they exchange" [1]. A COI consists of a collaborative group of users. Their members are responsible for making information:

Visible: discoverable by most users

Accessible: connected to the network with tools to use and provide assured access

Governable (institutionalized): governed with sustained leadership; institutionalize data approaches

Understandable: syntax (structure) and semantics (meaning) are well documented

Trusted: source authority (pedigree, security level, access control) known and available

These five components are essential to the data interoperability necessary for effective information sharing, publishing, and awareness training.

The COI concept is furthered described in 8320.02-G, "Guidance for Implementing Net-Centric Data Sharing," which discusses the roles, responsibilities, and relationships of the COI in information sharing. The primary responsibilities of COIs include:

Identify data assets and information-sharing capabilities to conform to data strategy.

Identify approaches and value measurement for the consumers of shared data.

Manage semantic and structural agreements to ensure that data assets can be understood and used effectively by COI members and unanticipated users.

Register appropriate metadata artifacts for use by the COI members and others.

Partner with a governing authority (GA) to ensure that COI recommendations are adopted and implemented through programs, processes, systems and organizations.

To enable trust, data assets shall have associated information assurance, security metadata, and an identified authoritative source when appropriate.

Reliable and Accurate Data for Subsequent Use by Customers

Not all data are created equal. Because many sites replicate data or provide repackaged data, it may be difficult to weed through various data sets to find the most current and most dependable data. Data from a reliable source (a source that is approved by the data producer) or from an accurate source (a source that can certify the content) is more in demand and effective for data reuse. Therefore, reliable and accurate data is an important attribute for an ADS. As the underwear bomber case revealed, because both the TSDB and TSA's No-Fly List database are functional combinations of data sources, the associated data are no longer accurate. Therefore, they cannot be certified as authoritative data sources.

Functional Combination of Multiple, Separate Data Sources

Many data collections do not come from authoritative or trusted sources. We recognized that unofficial sources create confusion among general data consumers by providing "non-maintained" duplicative data and "unofficial" parcel-like data sets that can unexpectedly harm or damage application users and developers with inaccurate, out-of-date information. For application services to be credible, they must pull their data from authoritative data sources.

Here we combine the five attributes listed above and propose two important concepts, Designated Data Source (DDS) and Authoritative Data Element (ADE) which will be essential to protect data consumers, and support data providers and application services in their authorized role of data stewards and to ensure that the user community has the best available and most current, reliable, and accurate data.

A data source designated by a governance authority is a Designated Data Source. It is used to perform special operations such as update monitoring and proactive pushing. In order to maintain data quality and consistency throughout a state's entire data reporting source system, each data element is required to have only one authoritative (primary) system source. The designated system is the only system allowed to report those data elements unless otherwise approved by the data governance authority. These data elements are called authoritative data elements (ADEs). Table 1 shows a set

Table 1. Authoritative Data Elements from a Guard Information Retrieval System

Designated Data Source	Purpose	Authoritative for
CorpDB	Foundation for Corp	Employee ID, First Name, Last Name, Corp Enterprise Username
CredentialDB	PKI Certificate Data/ Login Credentials	PKI Certificate
PayDB	Occupational and Demographic Data	Paygrade, Payplan
HRDB	Occupational and Demographic Data	Rank
ContractorDB	Contractor Validation System	
OrgContract	Location and Contact Data	Email, Address, Phone Number
HRContact	Location and Contact Data	Email, Address, Phone Number

of ADEs defined in a Guard Information Retrieval (GIR) System project with associated designated data sources.

The authoritative data elements listed in the third column of Table 1 must have the definitions, rights, pedigree, and data quality in the GIR project and that must be how it is described and presented to the public. Without that, it can cause considerable confusion about its value, ownership, and so on. These ADE data can be created and maintained by multiple, separate data sources across the country. Email, address, and phone number are three ADEs which have OrgContract and HRContract as data sources. Although a data element created within a data source has the original business purpose specific to a particular application service, over time it can evolve to become critical to other business operations because of the level of detail it provides.

With the five attributes defined in this section in mind, we therefore have an ADS defined as a data source with a designated mission statement whose products have undergone producer data verification, validation, and certification activities.

Applications of Authoritative Data Sources

In this section we illustrate how ADS was used by the National Spatial Data Infrastructure (NSDI) and can be used by the secure Identity Ecosystem coordinated by the National Strategy for Trusted Identities in Cyberspace (NSTIC) to be launched by a private-sector initiative as a nationally interoperable framework of independent federated identity systems. ADS can help improve privacy and security implementation. If it is done properly, the Identity Ecosystem will provide value to any participant who needs to verify a user's identity, and will provide tremendous opportunities to streamline the further commoditization of human identity. We first describe the National Cadastre within the NSDI to manage a single source of authoritative cadastral data that is controlled and managed by designated data stewards [5]. We then describe the Identity Ecosystem infrastructure plan to manage identity data through an ADS Framework and as strategic national assets—and protect them while safeguarding privacy and civil liberties. This project is a national security priority and an economic necessity.

NATIONAL SPATIAL DATA INFRASTRUCTURE (NSDI)

Digital cadastral data is a spatial representation of property boundaries and the related property description of land parcels. Digital data is provided in many ways to meet the different needs of customers. Example data include administrative area data, cadastral data, drainage data, elevation data, water data, and so on. Cadastral data defines the extent of rights in land, the characterization of those lands, and the status of ownership, and when presented to the public it may cause considerable confusion related to value, ownership, and so on. Cadastral data are also unique because they are created and maintained by over 4,000 separate entities across the country [2, 4]. As a result, there is a high demand for compiled and standardized cadastral information that will be incorporated into a wide variety of applications including emergency response and recovery, environmental management, health and safety, fleet management, energy management, and mortgage applications.

The vision for the NSDI and the National Cadastre within the NSDI is to have a single source of authoritative cadastral data that is controlled and managed by designated data stewards. The Federal Geographic Data

Committee (FGDC) Subcommittee for Cadastral Data has developed a series of documents over the past ten years that describe the concepts and polices related to the creation, use, and publication of cadastral data [5]. They not only help to characterize and clarify the subtle differences of terms, but their relationships as well:.

1. The relationship begins with the Legislation/Executive Order/ Ordinance. This is the authorization giving the official recognition of authority to create an ADS.
2. The ADS becomes the data steward and collects and maintains data. This data is then labeled the authoritative data.
3. Regular updates (at least annually) are conducted thus creating a trusted source which publishes and provides access to what now is labeled trusted data. The trusted source has an understanding agreement with the authoritative source to provide standardized data and metadata.
4. The final step is to provide access.

The ADS may also handle special requests which reside outside of those who have trusted source access.

Benefits of Authoritative Data Sources

A study result from the Energy Community and Cadastral Data FGDC Cadastral Subcommittee concluded that, from a regulatory perspective, anecdotal studies in Wyoming have established that a common survey-based Cadastral NSDI can save as much as 50 percent in application, permit, monitoring, and reclamation activities. In the cases examined, the time for the state regulatory agency to process an application for a permit to drill (APD) was cut in half because all required information could be easily cross-referenced and validated. The added benefit of the increased accuracy of information and more complete understanding of the data and special circumstances is immeasurable [3].

Value of Identifying Authoritative Data Elements

Many analyses and reports [3, 6] have indicated the importance and value of identifying authoritative data elements. To identify authoritative data elements without DDS, many users have to go through many data sources, which is time consuming for both users and sources. Not only do these tasks require many months and many people, but the pedigree and quality of the data sources are mostly unknown or ambiguous. That creates maintenance issues that are difficult and expensive for data consumers and application developers.

Identifying an authoritative data element with designated data sources allows data consumers to achieve immediate access to the best initial data product. These authoritative data elements are clearly described and they guarantee pedigree and data quality. They may have supporting configuration management tools for metadata, which not only can save knowledge acquisition time, but also can enhance data development and deployment.

NATIONAL IDENTITY ECOSYSTEM (NIE)

We use the Internet and other online environments to increase our productivity, as a platform for innovation, and as a venue in which to create new businesses [8]. The United States faces a host of increasingly sophisticated threats to sensitive and confidential personal and corporate information. Fraudulent transactions within banking, retail, and other sectors—along with online intrusions into the nation's critical infrastructure, such as electric utilities—are all too common. As more commercial and government services become available online, the amount of sensitive information transmitted over the Internet will increase. The digital infrastructure, therefore, is a strategic national asset, and protecting it—while safeguarding privacy and civil liberties—is a national security priority and an economic necessity. The President's Cyberspace Policy Review established trusted identities as a cornerstone of improved cyber security.

The Federal Identity, Credential, and Access Management (FICAM) effort addresses trusted digital identities. Its roadmap provides federal agencies with architecture and implementation guidelines to meet the identity, credential, and access management challenges they face every day. FICAM introduced the Authoritative Attribute Exchange Service (AAES) capability as the means of securely sharing authoritative identity attributes within an agency. To support the AAES capability, agencies must streamline the collection and sharing of digital identity data, establish an enterprise digital identity model, identify authoritative data sources, and streamline the processes used to populate those authoritative data sources. Like any high-volume enterprise with critical security and privacy demands, agencies need to secure complex heterogeneous environments with multiple identity sources and applications. To achieve this, administrators must manage the multitude of different authentication and authorization methods and requirements that are part of such a disparate environment.

The strategy on the National Identity Ecosystem (NIE) intends to accelerate those activities and extend trusted digital identities beyond the federal boundaries and into the national domain. It is designed to correlate identity attributes from the various authoritative data sources across agencies and private enterprises, and provide a single authoritative source of digital identity. This strategy focuses on transactions involving the private sector, individuals, and governments. It addresses the international nature of many transactions. It also recognizes ongoing public and private sector efforts relative to trusted identities, and builds on them for application in the larger national and global forum for online services. By themselves, trusted digital identities cannot solve all security issues associated with online transactions, but trusted digital identities do play a critical role in the overall enhancement of security in online services in a manner that promotes confidence, privacy, choice, and innovation.

The NIE provides a means of securely exchanging digital identity attributes between authoritative data sources and the organization's consuming applications. However, in many instances, this data is spread across multiple authoritative sources within the agency, thereby complicating the challenge of exchanging attributes between sources and consumers.

Both NSDI and NIE projects require well-managed verification, validation, and certification management processes for their Authoritative Data Sources infrastructure.

Conclusion

The cover story for the January 2012 issue of *Popular Mechanics* entitled "Digital Spies: How China's Secret War Threatens Our Economy, National Security—and You," made it evident that awareness of cyber threats has gone mainstream. Cyber security has taken on a new urgency in U.S.

federal government circles, leading to plans to create an Identity Ecosystem in an effort to assign identifying markers to all American Internet users. The plan—as drafted by the Obama administration—is known as the National Strategy for Trusted Identities in Cyberspace (NSTIC). The Identity Ecosystem is an online environment where individuals, organizations, services, and devices can trust each other because an ADS establishes and authenticates their digital identities. Today's cyber threats often use "outsourced" resources for data acquisition and storage, stealthy access to systems, identity authentication, identity collection and theft, and so on. Cyber security faces daunting problems related to authoritative data sources, electronic discovery, data quality, and anomaly detection. For example, we need an authoritative data source database of national vulnerabilities that supplies every known vulnerability in every known IT system and software application so we are not flying blind. An ADS will provide the ability to obtain timely, actionable, and reliable cross-functional readiness and resourcing information at a level suitable for manning, training, equipping, and stationing purposes.

The ability to maintain livable communities and manage authoritative data sources is becoming increasingly important for developing strategies to produce high accuracy data sources and generate actionable intelligence for organizational decision making. Data from multiple sources are integrated to meet a majority of the needs and are becoming commonplace in organizations charged with supporting a myriad of demands for actionable intelligence. Such data and technology significantly increase the capacity of a COI to respond to emergencies in a timely and effective manner. An ADS implementation can help to deliver the right information to the right person at the right time, and help organizations to deploy traditional security controls, processes, and leading-practice architectures. The cyber security infrastructure not only needs to ensure data quality, but also to find those trusted data sources that we need to operate and propagate as opposed to using data that is not authoritative but is rather derived data or enhanced data. We need to identify these ADSs, collect them, assign data stewards, validate their data quality, and make them available, accessible, and discoverable by the COI that needs them.

This chapter described the idea of establishing a formalized form for authoritative data sources that can be suitable for communication, interpretation, or processing either by human or by automated systems. We hope data quality [10] for ADS can be measured and managed to ensure reliable and accurate data for users.

REFERENCES

[1] Army Net-Centric Data Strategy. Authoritative Data Source (ADS). [Online]. Available: http://data.army.mil/datastrategy_authoritative.html

[2] Authority and Authoritative Sources: Clarification of Terms and Concepts for Cadastral Data Version 1.1. s.l.: FGDC Subcommittee for Cadastral Data, August 2008.

[3] B. Johnson, B. Ader, N. von Meyer, S. Kirkpatrick, M. Birtles, and W. Grayson. (2006, May). Cadastral Data and Energy Industry. [Online]. Available: http://www.nationalcad.org/showdocs.asp?docid=171&navsrc=Project

[4] Cadastral NSDI Reference Document. (2007, October). [Online]. Available: http://www.nationalcad.org/data/documents/Cadastral NSDI Reference Document v11.pdf

[5] FGDC Subcommittee for Cadastral Data. Authority and Authoritative Sources: Clarification of Terms and Concepts for Cadastral Data. [Online]. Available: http://www.nationalcad.org/data/documents/Authority and Authoritative Sources Final.pdf

[6] C. Hill and G. Jones, *Strategic Management*, Houghton Mifflin Company: New York, 2008, p. 11

[7] Wikipedia. Data Source. [Online]. Available: http://en.wikipedia.org/wiki/Data_source

[8] National Strategy for Trusted Identities in Cyberspace, April 2011, White House. [Online]. Available: http://www.nist.gov/nstic

[9] K. T. Huang. (2009, March 9). "On the Authoritative Data Sources: One Data Element at a Time." Presented at DAMA-NCR Chapter Meeting. [Online]. Available: http://www.dama-ncr.org/Library/2010-03-09-Kthuang-ADS-Talk-Taskco.pdf

[10] R. Wang. The MIT Total Data Quality Management Program. [Online]. Available: http://web.mit.edu/tdqm/www/about.shtml

[11] The BEACH Project. "Bettering the Evaluation and Care of Health." Conducted by the Australian General Practice Statistics and Classification Centre (AGPSCC). [Online]. Available: http://sydney.edu.au/medicine/fmrc/beach

The Evolving Consumer Online Threat Landscape: Creating an Effective Response

Adam Palmer

The proliferation of Internet-connected devices in the consumer market has created a dramatic shift from a single point PC-based threat risk to an expanded threat perimeter that includes tablet devices, smartphones, and Internet-connected televisions. Consumer home network growth now requires security for a consumer's entire digital lifestyle and not just a single-point stationary device. The difficulty of securing this expanded threat landscape is exacerbated by the rise of unique attacks that sometimes render traditional antivirus strategies ineffective. This expansion of the consumer digital threat landscape presents new challenges for law enforcement and security professionals. The solution is improved reputation-based threat detection, effective public-private collaboration and increased user responsibility within the digital ecosystem.

The Expanding Consumer Home Network

Evidence of the expanded threat against the digital home network was most recently demonstrated by the detection of the psybot botnet, regarded as the first recognized attack capable of directly infecting home routers and cable/ DSL modems. First observed infecting a Netcomm NB5 modem/router in Australia, researchers concluded that the psybot (or Network Bluepill) botnet was merely a test virus designed by cybercriminals to confirm the capability of a home network attack strategy. The most recently discovered version of this attack tool targets a wider range of devices, and contains the shellcode for over fifty popular models of cable and DSL modems. The exploit is very difficult to detect; the only way to discover it is to monitor traffic from the router itself, which is beyond the reach of ordinary desktop security software. By attacking the expanded home network, the cybercriminal gains the advantage of attacking entry-point platforms that make detection difficult and that provide the malicious code with multiple potential infection points.

An additional example of the expanding consumer digital threat landscape is the proliferation of smartphone devices. While open operating platforms provide developers the freedom to innovate, they can also be misused by malicious developers to create applications designed to distribute malware. Although malicious smartphone applications are not pervasive, evidence has been detected that cybercriminals are experimenting with open platform mobile applications as a distribution channel for malware. The proliferation of open developer applications for mobile connected devices used by consumers is likely to continue to attract cybercriminal activity across these new distribution channels.

Much has been said about predicting threats to mobile devices. A perhaps often-overlooked solution to mobile security may be found in remembering that these devices are "mobile." In the past, many mobile security efforts have failed because they misunderstood the mobile user's main concern: Mobile devices can be easily lost in the physical world. This is not just a cyber threat—it is a physical danger. Any successful new mobile security effort for users has to provide users the capability to deny control to a thief who has stolen the device. After the physical device is secured, the main issue then becomes educating the public about the potential dangers of having unsecured data on a mobile device. There is currently a sense among some users that mobile devices are not as vulnerable to attack as a traditional PC.

As noted above, new open platforms prove the reality is more complicated and is still developing. Any open system is also an open target for cyber-crime. The challenge is to provide a security tool that can be used without causing performance disruption on the device because users are more likely to allow security that does not disrupt performance. This means that any mobile security tool has to be fast and cannot disrupt other applications. Because mobile devices have limited capacity compared with the typical PC, mobile security must be more sensitive not to waste precious memory.

Finally, to improve mobile security, software and hardware developers need to prioritize security in the development stages of new devices. Too often there is a rush to launch a new technology, with security being an afterthought. This attitude must change. Achieving mobile security will depend on industry adoption of security standards, cooperation with law enforcement, and a clear understanding of the different requirements between PC and mobile security.

Utilizing Reputation-Based Security to Meet the Challenges of the Expanding Threat Landscape

In addition to attacking multiple consumer devices beyond the traditional PC, cybercriminals are also distributing millions of distinct threats intended to infect each device with an unknown threat variant. This paradigm shift is a huge challenge for traditional fingerprint and heuristic-based antivirus techniques. Given the rarity of each threat, many threats are likely to never be discovered, and if they're not discovered they can't be fingerprinted. It is also inefficient to release a fingerprint to hundreds of millions of users when that fingerprint might only protect one or two users across the globe. With hundreds of thousands of new malware variants being released, it is increasingly difficult to create, test, and distribute the volume of signatures necessary to address the problem.

A solution to the evolving threat landscape of unique variant viruses is to reinvent an antivirus and develop an entirely new approach that accurately classifies files based on their distribution (or lack thereof). Symantec has utilized over 58 million participating users who voluntarily submit anonymous, real-time telemetry data about the applications they use. In addition, Symantec leverages data from its Global Intelligence Network, various web crawlers, other Symantec product offerings that have data submission capabilities, Symantec's Security Response organization, and legitimate software vendors who provide application instances to Symantec. This data is incorporated into a massive model that represents executable files (and associated file metadata), anonymous users, and the linkages between them. This data is then collected into reputation algorithms to compute highly accurate reputation ratings on every single file, both good and bad, known to each participating user. Such an approach is not only effective for detecting popular malware, but especially those affecting just a handful of users across the entire Internet. The system doesn't just identify and classify bad files; it computes a reputation score for every single executable file, both the bad ones and the good ones, used by every participating user.

A file's reputation score is really a measure of whether its track record (when viewed using telemetry data from a large user base) is consistent with that of legitimate applications. For the most part, a file's reputation score will coincide with whether or not that file represents a threat to the end user (e.g., many threats are extensively polymorphic, causing them to have no real track record whereas many legitimate applications are found on numerous machines). In many enterprise situations, this distinction is less material. In particular, a typical IT security manager might block access to any applications that have not built up an appropriate track record (as long as all the bad applications are blocked). While blocking some lesser known, but still legitimate, applications will be a byproduct of this policy, the trade-off may be well worth it: malware (especially lesser known targeted malware) can be blocked as well.

Overall, reputation-based security provides the greatest malware detection capability and improves threat detection standards for both business and consumer users across all connected devices on which the system is deployed.

Norton Cybersecurity Institute: A Model Solution for Public-Private Collaboration to Stop Cybercrime

Although tools like reputation-based security software for Internet users are important, complete online protection can only be achieved by supporting law enforcement efforts to eliminate cybercrime. To further this goal,

Symantec has recently established the Norton Cybersecurity Institute (hereafter the "Institute"). Its mission is to facilitate training and collaboration with private industry and government to eliminate cybercrime. This is a unique collaborative model of global industry, law enforcement, and consumer safety groups. Collaboration among these groups contributes to a more timely understanding of the most significant cyber threats and increases consumer cybercrime awareness.

The Institute's core principles are:

Global collaboration is critical to reducing cybercrime
Internet safety can be improved through good security practices
Cybercriminals can be defeated

Institute-sponsored initiatives are designed to improve anti-cybercrime capability through technology, training, and strategy innovation. Programs are designed to assist law enforcement at the "tip of the spear" against cybercrime. By collaborating with Symantec, law enforcement receives the benefit of Symantec's technical and practical knowledge base to stop cybercrime. The result is improved safety for Internet users and the fulfillment of the promise of a safe Internet experience. The Institute is driven by the belief that the Internet is a great resource and that cybercriminals can be defeated. Cyber security can be achieved by effective Internet user education, good security practices, and effective law enforcement action to stop cybercriminals.

The Consumer Threat Landscape

When you use a desktop, smartphone, or tablet computer, it's easy to forget that the individual device is part of a network ecosystem. The illusion of individuality on the Internet creates the false perception that individual actions will not impact the larger network. This is, of course, not reality. The Internet is an ecosystem. One infected part affects the health and functioning of the entire body.

Securing the Internet cannot be outsourced, ignored, or forgotten. It is the responsibility of every user. We live in a connected world in which industry, government, and consumers share the same communications channels.

In essence, we are all in this together. Because one person's lack of responsibility not only harms that individual but provides a platform for other innocent users to be attacked, understanding consumer attitudes toward security is critical to improving security. Each year Symantec produces a threat report that provides a clear picture of consumer's attitudes toward the online threats.

In Symantec's "2010 Norton Cybercrime Report: The Human Impact," surveying more than 7,000 adults in fourteen countries, four clear themes emerged from the data [1]. All of the data concerned the human impact of cybercrime, with 65 percent of respondents reporting that they had been victimized by cybercrime. The four themes that emerged indicated that:

> Cybercrime is a silent epidemic, the result of a silent majority of people who feel powerless against cybercrime—only 3 percent think they will not be victimized.
> People reported feeling angry and victimized—but they are at a complete loss about what to do next because they believe that criminals will not be brought to justice.
> There is a large grey area when it comes to morals and ethics on the Internet.
> People are trying to protect themselves but are coming up short.

Despite the fact that there is a significant amount of unethical and even illegal behavior occurring online, only a fifth of consumer survey respondents expressed regret for their unethical behaviors online. Most notably:

> Eleven percent think it's OK to impersonate someone online.
> Twelve percent think it's OK to use someone else's research.
> Twelve percent think it's OK to browse someone else's files and email.
> Five percent think it's OK to hack into someone's computer and sell their personal information online.

In balance, the good news is that most people feel it is unethical to hack into someone's computer (87 percent), impersonate someone online (69 percent) and sell someone's personal information (87 percent). However, it appears that there remains a significant need for educational awareness about ethical and legal Internet user behavior.

A Call for Responsibility in Cyberspace

In many areas of our lives, we have learned to take personal responsibility and have come to understand the effect of our actions on others with whom we share the same environment. People are aware of the impact of pollution on the environment. Individuals who carelessly pollute experience well-deserved community disapproval. Their irresponsibility causes problems in the ecosystem that we all share. As we enjoy the Internet and appreciate the benefits it brings to our lives, awareness has to be raised that this is also part of the sensitive environment we all share. Security starts with each individual user's responsible behavior in the cyber world. We will have a much safer cyberspace if those users who ignore security feel the same disapproval that we apply to irresponsible behavior in the physical world. Keeping cyberspace safe starts with improving the individual security awareness of all Internet users.

Conclusion

Every user should be able to enjoy the Internet without fear of victimization. Empowerment will occur by raising awareness of the issues related to cybercrime and educating people on best practices and technologies to prevent becoming a victim. We are each capable of contributing to the reduction of cybercrime worldwide. The use of technology in our lives is a great benefit, but unsecured digital devices can also cause significant harm. By raising consumer awareness of digital security and reducing unsecured devices available for criminal activity, the global community will become safer for everyone to enjoy.

REFERENCES

[1] Norton Cybercrime Human Impact Report 2010.

Partners in Cybercrime

Eileen Monsma, Vincent Buskens, Melvin Soudijn, and Paul Nieuwbeerta

Hijacked online banking sessions, theft of credit card data, virus infections, spam . . . living in the age of the world wide web implies that we are all vulnerable to cybercrime—crime committed using mainly computerized means [21]. In the 1990s, cybercrime primarily seemed to be the domain of computer savvy youngsters motivated by recognition for their skills [27]. However, the rapid development of computer technology in recent years has been immediately followed by an increase in cybercrime and associated costs. In line with the "criminal enterprise model," cybercriminals are now seen as profit-seeking entrepreneurs who follow ordinary market rationality: they obtain profits by meeting a demand for illegal goods and services [10]. Cybercriminals barely have physical contact, but meet each other in online forums [13]. Some forums are open for anyone who is interested in this sort of illicit activity, whereas more professional forums operate underground and require good reputation and connections in the cybercrime scene.

One specific type of cybercrime that has grown substantially over the past years and has become the focus of many cybercriminal networks is trade in stolen personal financial information such as credit card data and login credentials for online banking that can be misused to make money [18]. In the cybercrime underground, the process by which data are stolen, resold, and ultimately used to commit fraud is known as carding [23].

The body of scientific literature in the field of cybercrime studies in general and carding in particular is rather limited. There are some descriptive studies exploring the market for stolen data and involving modus operandi [11, 13, 19]. Yet, because most researchers do not have access to sufficient data, there is a lack of knowledge about network structures and the identification of important individuals within these structures. This chapter, however, uses unique data that allow for network analysis. The data source is a digital copy of one of the biggest underground carding forums operating today. It includes all communications between both English and Russian speaking members (1,856 members) over a time span of about five years, as well as member profiles with individual characteristics. The main section of this forum consists of different subforums where members can participate in discussions and exchange knowledge on carding activities. There is also a section where some of the members are allowed to offer their goods and services. This is comparable to a regular online marketplace, such as eBay, except that all offers are related to carding. Furthermore, members can contact each other privately to set up a deal using the forum's instant messenger system. Hence, the relationships formed using that system indicate who attempts to do business with whom.

This research looks at cybercrime by analyzing social networks used by cybercriminals. Because a social network is defined as a set of actors (usually individuals) connected by ties (usually the personal relationships between them) [28], a cybercrime network is defined as a set of cybercriminals and their collaborative relationships. Centrality measures have been widely used to quantify the importance of individual actors in networks [28]. Therefore, this research aims to describe differences in individual centrality in the cybercrime network under study and investigate explanations for these variations. Based on social capital theory, it is expected that individuals select partners in crime on the basis of their resources, reputation, and status [6] and thus these characteristics are likely to be related to central network positions. Therefore, the research questions are formulated as follows: "To what extent are there differences in individual centrality in a cybercrime network, and can these differences be explained by resources, reputation, and status?"

This study contributes to existing literature in three ways. First, whereas most empirical studies on cybercrime or physical criminal networks are not theory driven, we take advantage of the extended body of literature in the field of social networks and social capital. Theories developed in this literature, combined with insights from law enforcement agencies dealing with cybercrime, provide a theoretical framework for our research. Second, we use uniquely rich data, which enable us to bridge the gap between criminological and social network literature. Criminologists studying criminal networks usually do not have large samples and sufficient relational data. As a consequence, they refer to the concept criminal network, but fail to evaluate network characteristics and empirically test relationships between individual attributes and network positions. Even in social network research, it is rare to have direct measures for relations because of practical and privacy reasons. In the current study, social network analysis will be conducted to describe differences in individual centrality scores. Logistic regression will be used to examine the effects of resources, reputation and status on the likelihood of the existence of a relationship and as such, explain variation in centrality. Third, this study does not only fill a theoretical and methodological gap within scientific literature, but evaluating cybercrime networks from a social network perspective can also lead to new insights for law enforcement. Knowledge on differences between and determinants of individual positions within a cybercrime network might help to develop more effective strategies to combat this type of crime.

Carding

To give an impression of the context under study, we start by describing what carding actually is. The term *carding* refers to the assortment of activities involved in the process of data theft, trade, and fraud with financial assets [23]. The data are personal financial information such as data stored on credit and debit cards and login credentials for online banking that can be misused by criminals to make money.

The carding process starts with data theft. *Hacking* is a method that is often used to obtain large volumes of data. Hackers remotely access computer systems of (online) business and financial institutions in order to steal their electronic files with sensitive customer information, or they install spyware on personal computers to log all usable data. Another method is *skimming*, which refers to techniques for copying data from physical cards when they're inserted in automatic teller machines (ATMs) or point-of-sale terminals (POSs). Another data theft technique is *phishing*, a method to gather personal information using seemingly trustworthy content such as faked emails or websites asking bank customers for their login credentials. These are just a few examples of how cybercriminals can obtain personal financial information [22].

Data thieves seldom seem to use the data themselves [22]. Carding forums, such as the forum under study, facilitate online trading in stolen data. In the forum's market area, data thieves and resellers offer the information for sale, usually in bulk. So-called carders will then buy the data to turn it into money. Cashing out the data is quite complicated, because the money has to cross both national and monetary borders and the cash flow to the carder needs to be covered. For this reason, carders typically make use of *cashers* who, for a fee, carry out the necessary fraudulent activities.

The most common types of fraud are online carding, in-store carding, and cashing [22, 23]. The first method, online carding, refers to ordering goods on the Internet using stolen credit card data. To avoid detection, the goods are sent to middle-men (called *drops*), who immediately forward them. Subsequently, the goods can be monetized by selling them on the black market. The second method, in-store carding, refers to purchasing goods in physical stores using counterfeit cards encoded with stolen account information. This is more complicated and risky than online carding, because it requires criminals to actually go to brick-and-mortar stores. The third method, cashing, refers to techniques that directly cash-out funds from the compromised financial accounts—for example, using counterfeit cards with the corresponding pin to obtain cash at ATMs, or through fraudulent bank or money transfers. The latter requires intermediaries (*bank drops* or *money mules*) who are usually recruited via spam and, for a fee, provide their bank accounts for deposits. They withdraw the received money from their accounts and use money transfer services to send it to other money mules in the carder's country.

All these illicit online activities are facilitated by security and web services that make it hard to trace back real identities. Cybercriminals need to make use of services that facilitate anonymous surfing and communications on the Internet. Furthermore, they need servers that are secure against criminal investigations to, for example, store gathered data, send out spam, or host illicit websites [22]. This whole chain of activities reveals that the carding process requires a wide variety of techniques and as such, many individuals with specific specializations are involved in the underground economy of carding [22, 27].

Centrality in Cybercrime Networks

The instant messenger system of the cybercrime forum under study enables members to privately contact potential business partners. The set of members connected by these personal relationships can be considered a social network. Characteristics of individual positions within such a network can be evaluated. Specifically, centrality is often used to quantify the importance of actors [28]. Commonly, different categories of centrality are distinguished [12], such as betweenness, outdegree, and indegree. Betweenness indicates how often a person is on the shortest path between two other actors. Being located between two others is usually considered a central and important position because it enables one to control communication. Degree refers to the number of relations a person has. Because one actor in the network actively seeks contact with another actor, the ties in the network are *directed* from the initiating actor to the other actor. As a result of this, indegree and outdegree can be differentiated. Members who actively contact many others by sending them private messages have many outgoing ties (i.e., a high outdegree). This is an indicator of their communication activity. Members who are contacted by many others have many incoming ties (i.e., a high indegree). This indicates their popularity as contact persons.

Typically, physical criminal networks are found to have a small core group of leaders who are able to facilitate and/or coordinate the different subtasks of a crime and, as such, connect otherwise unconnected peripheral criminals and occupy the central network positions [16, 17, 26]. The actors in the network under study, however, have virtual instead of physical relationships. Scientists argue that virtual networks differ from physical networks. The main reason is that virtual ties require less investment and thus are less costly, which makes it easier to extend one's personal network without the help of others [14]. Forum members do not need to put much effort into finding and connecting to others, because the forum provides a clear overview of the resources and needs of members and anyone can contact anyone else using the instant messenger system. Based on this, it is expected that cybercriminals on forums do not need leaders to centrally organize cooperation, because they can easily connect to relevant partners themselves. The situation probably can be characterized as self-organization; that is, actors self-organize network structures by interacting locally and without central direction [13]. This does not exclude the possibility that the network might still be centralized to some extent. In the theoretical framework discussed in this section, possible explanations for differences in individual centrality in a cybercrime network will be presented.

RESOURCES

Social capital theory gives more insight into the reason people form networks by connecting to each other. As Coleman explains, actors need resources to achieve their goals [6]. This gives people an incentive to select relations on the basis of personal gains [5], which implies that individuals who need the resources of many others to pursue their interests are likely to actively get involved in many relations and thus have a high outdegree. People who control resources that interest many others are popular relations and are thus likely to have a high indegree. People who are minimally interested in others' resources and have little control of resources others are interested in are likely to occupy peripheral positions.

Coleman's arguments can be applied to cybercrime networks. Findings from criminal investigation reveal that most cybercriminals seem to choose one specialization [6, 19, 22]. Therefore, they need to exchange resources to make their activities profitable. For example, one member offers security and web services (tools to facilitate and cover illicit online activities). This enables a data thief to steal and store personal financial information. The data thief then sells the data to a carder. Another party offers cash services to that carder, so that the carder can cash the value of the stolen accounts he bought. In line with Coleman's arguments, the system is characterized by interdependence [6]. Specifically carders need to be involved in many ties to pursue their interests. Therefore, they are likely to actively contact many others and they are relatively more often on the shortest path between two individuals, namely between a data thief and people who offer cash services. This results in the hypotheses that *carders have a higher betweenness in the network than other members* (H1) and *carders have a higher outdegree in the network than other members, that is, they send more messages* (H2).

At first glance, it seems that carders have the same position as leaders in physical criminal networks. They act as brokers by occupying the structural holes between data theft and cash services. Because a broker can control the chain of activities and make himself essential for the different parties, it usually is a very powerful and profitable position [1, 2]. In the context under study, however, every member can easily observe the structural holes between data theft and cash services on the forum, and the private messenger system allows everyone to bridge the observed holes by connecting to both parties. Buskens and Van de Rijt predict that if every actor strives for brokerage, the structural advantages disappear [5]. For this reason, the position of carders is expected to be less exclusive, and thus is assumed to be less powerful compared to the broker position of leaders in most physical criminal networks. In fact, carders do not possess interesting resources themselves and their partners are essential for them rather than the other way around.

There is also a supply-side to this underground economy. The forum administrators allow some members to advertise and sell products and services that are of crucial importance for trade in stolen data, namely the data itself, cash services and security and web services [19, 21, 22]. Many forum members are likely to be interested in these resources and it is therefore expected that the more often a member offers one of these services, the more popular the member becomes. This results in the hypothesis that *the more often a forum member offers services, the higher a forum member's indegree in the network, that is, the more messages a forum member receives* (H3).

REPUTATION

Coleman adds to his arguments on the need to exchange resources that transactions usually involve a risk because resources need to be invested some time before returns are received [6]. For example, somebody who wants to buy stolen data might need to pay before receiving the data. The buyer risks receiving useless data or no data at all. Therefore, people do not only select relations on the basis of resources, but also on the basis of trust [6]. According to Buskens and Raub, the placement of trust depends on the mechanisms of learning and control [4]. Learning refers to the extent to which an individual can benefit from information about the trustworthiness of a potential partner. Control refers to the extent to which an individual can sanction or reward a potential partner after a transaction. At the dyadic level, an individual can learn from past experiences with a certain partner and control that partner's behavior by threatening not to cooperate in the future if trust is abused. On the forum under study, however, members often initiate relationships with strangers and possibly only for a single transaction. In these cases, potential partners do not have a shared past and future to rely on. Nevertheless, both individuals are embedded in the same network. At the network level, individuals can learn from information about a potential partner that is circulating in the network (i.e., reputation). They can also control the other's behavior through the possibility of spreading information and affecting that person's reputation [24]. As such, reputation reduces uncertainty in exchange relations [7].

If people do not have physical but only virtual relationships, traditional means to share information related to reputation are missing and electronic substitutes are needed [7, 8]. A frequently used online alternative is a so-called feedback system that can be used to rate others. The aggregated ratings result in a reputation score that can assist others in selecting relations [7, 8]. Such a system is also used on the forum under study. Whereas in physical environments the diffusion of information related to reputation is constrained to existing relations in the network, an online feedback system provides global information; every individual can review every other's current reputation score [15]. In such efficient systems, reputation effects on cooperation are expected to be strong [24, 29]. Forum members with a better reputation are likely more popular business partners and thus are contacted by more members. This results in the hypothesis that *the better a forum member's reputation, the higher a forum member's indegree in the network, that is, the more messages a forum member receives* (H4).

STATUS

As described in the preceding section, reputation is based on ratings by other members. Another system that has been developed to signal the trustworthiness of potential partners is the status assigned by administrators, who function as the forum's governing council. An individual can learn from the signals administrators give by assigning a certain status. Moreover, the status system can also serve as a control mechanism. Administrators can be informed if a partner abuses trust so that the partner gets a negative status. For example, on the forum under study there are negative terms indicating whether someone has a certain "unresolved problem," broke the forum rules (referred to as a *deer*), cheated (referred to as a *ripper*), or in the worst case, is *banned*. As a consequence, others will probably avoid future transactions with that person. Besides signaling trustworthiness, a higher status also provides a member with forum-specific resources. Newcomers only have limited access to the forum. They can become official members if they have at least two good references, but that does not allow them to offer services yet. If the forum administrators believe that members are able to offer interesting and reliable resources, they get a *service* status and are allowed to offer services. Members who are known and trusted by the administrator gain a *verified member* status and members with a *moderator* status even have the right to supervise—and if necessary, intervene—in discussions on certain subforums. These forum-specific resources might also affect a person's popularity as a contact person. Although network positions are typically used to explain why some people have a competitive advantage and consequently gain a higher status [3, 20], this reasoning implies that the relationship between individual centrality and forum status is the other way around. People are likely to use status information to select relations and therefore individuals with a higher status are expected to become more popular partners, resulting in the hypothesis that *the higher a forum member's status, the higher a forum member's indegree in the network, that is, the more messages a forum member receives* (H5).

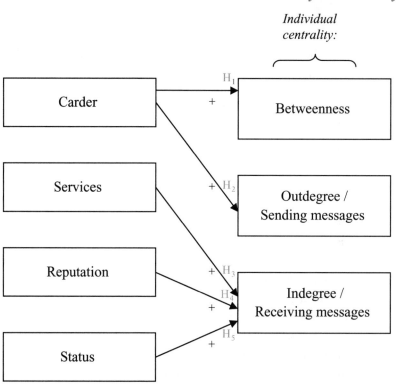

Figure 1. Hypothesized effects on individual centrality.

The hypotheses that have been formulated in this theoretical framework are summarized in Figure 1.

Data and Methods

In this section we explain our data source, then we briefly explain the methods used to determine the three common measures of centrality: betweenness, outdegree and indegree. Dependent and independent variables are incorporated into the data analysis in order to achieve the final results.

DATA

The study on which this chapter is based uses a unique data source—a digital copy of an underground online carding forum made by the Dutch National High Tech Crime Unit. The data include all forum postings (153,936 postings) and private messages (60,437 messages) sent by the 1,846 members from the start of the forum (early 2003) until the date the digital copy was made (the end of 2008) as well as member profiles with several individual characteristics, such as date of registration, feedback score, and status. The private messages sent between members indicate who attempts to start a business relation with whom and are used to construct a network with directed ties. As a consequence, only members who have used the private messenger system at least once during the complete time span are included in the analysis (N = 1,700). After the first contact, private conversations often seem to be continued using online messenger systems outside the forum environment. For this reason, the frequency of contact is not very meaningful, and dichotomous ties rather than valued ties are used.

To test the hypotheses, we do a cross-sectional analysis with the network over the complete time span. We construct a data set at the dyadic level including all possible relationships within the network (1,700 times 1,699 = 2,888,300 cases). These dyads are nested in senders and receivers, and a binary response variable indicates whether or not the tie between a sender and receiver actually exists. The relevant individual characteristics of the two actors involved in the tie are also included in this data set.

MEASURES

To describe differences in individual centrality within the network, three common measures of centrality are used [12, 28]. First, *betweenness* is measured as the percentage of all shortest paths between pairs of others that include the actor. Theoretically, the scores can range from 0 percent (the focal actor is never on the shortest paths between pairs of others) to 100 percent (the focal actor is on every shortest path between pairs of others). Second, *outdegree* is measured as the number of outgoing ties (i.e., how many others are contacted by the focal actor). Third, *indegree* is measured as the number of ingoing ties (i.e., by how many others is the focal actor contacted).

Dependent Variables

To test the hypothesis on betweenness, betweenness is measured at the individual level, resulting in the same measure as described above: the percentage of all shortest paths between pairs of others that include the actor. For the hypotheses on indegree and outdegree, however, the dependent variable is measured at the dyadic level. We use a binary variable, indicating whether or not (1/0) there is a tie between a sender and a receiver. The probability that a sender actually has an outgoing tie is larger for a sender with a higher outdegree, and the probability that a receiver has an ingoing tie is larger for a receiver with a higher indegree. In the "Analytic Approach" section later in this chapter, we describe the analytic approach and explain why indegree and outdegree are not measured at the individual level but by one binary variable at the dyadic level.

A limitation of the data is that private messages that are deleted by the sender, receiver, or administrators are not traceable. There is also a chance that some relations are not included at all because some members refer to their e-mail or a messenger system outside the forum environment for a reply on their forum posting. As a consequence, degree centrality scores might be underestimated. Furthermore, the virtual identities, that is, nicknames, do not necessarily correspond with unique physical identities. Individuals might share nicknames or have more than one nickname. Nevertheless, they are treated as separate and unique members on the forum and therefore are also considered to be unique actors in this study.

Independent Variables

The carder and services variables are constructed using forum postings. Specifically messages on the marketplace section, where members offer goods and services, are examined. To indicate whether a member is a *carder*, a dummy variable is created and coded 1 if the actor responds on subforums with data offers or on subforums with cash service offers.

Members can only start new threads on the marketplace section to offer *services*. Therefore, starts of threads on subforums for data, cash services, or security and web services (such as anonymous surfing or secure hosting) indicate how often a member offers one of these services (i.e., the services variable).

Table 1: Descriptive Statistics of the Variables

	Mean	SD	Range
Dependent variable at dyadic level (N=2,888,300)			
Tie	0.009		0/1
Independent variables at individual level (N=1,700)			
Carder	0.22		0/1
Services	0.08	0.40	0–5
Reputation	1.50	3.74	–2–40
Status	5.38	1.63	1–10

The forum facilitates a feedback system where members can rate each other. The aggregation of positive ratings (+1 for every positive rating) and negative ratings (–1 for every negative rating) results in an individual feedback score presented in member profiles; these scores are used to indicate *reputation.* Although people are more likely to rate others with whom they have a personal relationship, there is no predetermined correlation between reputation and degree scores, as ratings can be positive but also negative. Moreover, ratings do not have to be based on private contact but can also be based on a member's contribution to the accumulation of knowledge on discussion subforums.

The *status* of actors is drawn from member profiles resulting in an ordinal status scale from 1 to 10 (Banned user=1, Ripper=2, Deer=3, Unresolved problem=4, Newcomer=5, Member=6, Service=7, Verified member=8, Moderator=9, Administrator=10). The order of rank is based on information in the forum's guidelines.

Indegree and outdegree are found to be highly correlated (r=.856, p<.001), probably because ties are likely reciprocated. In order to examine the hypothesized effects over and above the effect of the existence of a *reversed tie,* we add a control variable indicating whether or not a tie exists the other way around (1/0). This variable actually mirrors the dependent variable *tie,* and thus has the same descriptive statistics. Table 1 shows descriptive statistics of the variables used. This table illustrates that a proportion of .009 (=0.9 percent) of the possible ties actually exists.

Pajek (Batagelj and Mrvar, 1996), a program for large network analysis, is used to visualize a network and estimate centrality scores. Based on these estimates, we can provide descriptive statistics and use them to describe differences in individual centrality. In order to test the hypotheses, two separate analyses are done. First, the effect of being a carder on betweenness is tested. In view of the skewed distributions and outliers of the variable, the Mann-Whitney U-test is considered the best option for this purpose.

The other hypotheses are tested using logistic regression. Logistic regression is a form of regression analysis that can be used to examine the effects of several predictors on the probability of the occurrence of an event. In our case, the event is the actual existence of a directed tie from a sender to a receiver. Hence, we perform dyadic-level analysis. What we test is whether the existence of a tie depends on characteristics of the sender (namely being a carder) and of the receiver (namely services, reputation, and status). Because having a higher probability of an ingoing tie is the same as having a higher indegree, and having a higher probability of an outgoing tie is the same as having a higher outdegree, this enables us to test the formulated hypotheses. For example, if a tie is more likely to exist if a sender is a carder, this means that carders have a higher probability of an outgoing tie and thus have a higher outdegree.

We have to take into account that dyads are nested in both senders and receivers and thus observations are not independent. If a sender sends a private message to a receiver, that does not only increase the sender's outdegree, but automatically also the receiver's indegree. Interdependency of observations violates the assumption of standard regression analysis and would result in biased standard errors [25]. Logistic regression with two-way clustering allows us to account for this property of the data by adapting the standard errors to correct for clustering in senders and receivers (Cameron et al., 2006) [9]. For the hypothesis on betweenness, we cannot account for interdependency of observations and we are aware that the significance of this effect might be overestimated.

Results

The results from the data analysis provide clear evidence of the differences in the centrality measures of indegree, outdegree and betweenness. By statistically linking data with a social network perspective, these results provide a keen insight to assist law enforcement understanding on how to target cybercriminals.

DIFFERENCES IN INDIVIDUAL CENTRALITY

In Table 2, descriptive statistics of indegree, outdegree, and betweenness are presented for different time intervals as well as for the entire period. Members are included in the network from the time interval in which they were registered on the forum until the end of the final period. The network size increases from 883 to 1,700 members over time. The scores give us an impression of the evolution of the network in terms of centrality. Table 3 shows that the mean degree increases over time from 1.14 to 7.50 new relationships. This indicates that members establish more and more new relationships. Over the complete time span, on average, individuals are connected to 15.75 others. The mean betweenness actually decreases over time, except for the peak at T_2 caused by an extreme value of 12.9 percent of all shortest paths between pairs that are found to include one certain individual. Over the complete time span, the mean betweenness is 0.10. This means that, on average, forum members are on only 0.1 percent of all shortest paths between pairs of others. It appears that actors are more likely to connect to relevant partners themselves than to communicate via others.

The range and the relatively large standard deviations of indegree, outdegree, and betweenness, as presented in Table 3, reveal that there is considerable variation in centrality scores. Figure 2 shows the in- and outdegree distributions for the complete time span on a logarithmic scale. Although a fair number of forum members occupy more peripheral positions, there are also many actors with relatively high indegree scores and many actors with relatively high outdegree scores. In contrast to "typical" criminal networks with only a small core group of central leaders and all other actors remaining in the periphery, the network under study consists of many well-connected members. Figure 3 illustrates that the core is actually so extended that the center and periphery can hardly be distinguished. Many members

Table 2: Descriptive Statistics of Indegree, Outdegree and Betweenness for Different Time Intervals

					Time interval					Complete time span (N=1,700)	
	T1 (2003–2005) (N=883)		T2 (2006) (N=1,211)		T3 (2007) (N=1,531)		T4 (2008) (N=1,700)				
Variables	Mean (SD)	Range	Mean (SD)	Range	Mean (SD)	Range	Mean (SD)	Range	Mean (SD)	Range	
Indegree	1.14 (3.10)	0–70	4.36 (12.01)	0–361	5.59 (11.47)	0–176	7.50 (15.56)	0–322	15.75 (28.11)	0–741	
Outdegree	1.14 (1.86)	0–16	4.36 (5.09)	0–53	5.59 (8.65)	0–94	7.50 (13.56)	0–223	15.75 (19.77)	0–309	
Betweenness	0.08 (0.29)	0–4.81	0.16 (0.55)	0–12.89	0.07 (0.30)	0–6.60	0.04 (0.21)	0–7.32	0.10 (0.63)	0–23.46	

Table 3: Logistic Regression (with Two-Way Clustering) of the Likelihood of a Directed Tie from a Sender to a Receiver on Characteristics of Both Sender and Receiver

Variables	Model 1	
	Coeff.(s.e.)	Marg. eff.(s.e.)
Carder (sender)	0.40(0.05)***	0.0028(0.0004)***
Services (receiver)	0.17(0.08)*	0.0011(0.0005)*
Reputation (receiver)	0.07(0.01)***	0.0004(0.0001)***
Status (receiver)	0.14(0.05)**	0.0009(0.0003)**
Reputation* Services (receiver)		
Status* Services (receiver)		
Reversed Tie	5.30(0.06)***	0.4454(0.0113)***
Constant	−5.59(0.05)***	
Baseline Probability of Tie		0.0093
Number of Ties	2,888,300	
Number of Individuals	1,700	
Log Pseudolikelihood	−98,495.41	
Df	641	

*significant at .05-level
**significant at .01-level
***significant at .001-level (one-sided)

occupy central network positions and thus, these positions cannot be considered exclusive.

CAN DIFFERENCES IN INDIVIDUAL CENTRALITY BE EXPLAINED BY RESOURCES, REPUTATION, AND STATUS?

The descriptive results presented in the previous section show that, although the central core of the network is quite extended, forum members do vary in the total number of relations and *in between* positions. This section examines which individual characteristics are related to central network positions and thus can explain the variation in individual centrality. Recall that for the explanatory analysis we do a cross-sectional analysis with the relationships formed over the complete time span, because crucial individual characteristics are measured only once.

We start with testing the first hypothesis: carders have a higher betweenness in the network than other members. Table 3 shows that the mean betweenness is 0.10—that is, on average, forum members are on 0.1

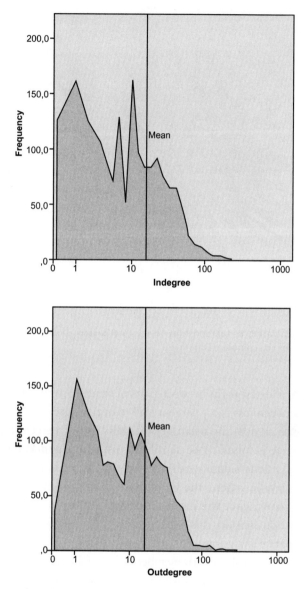

Figure 2. In- and outdegree distributions for the complete time span on a logarithmic scale.

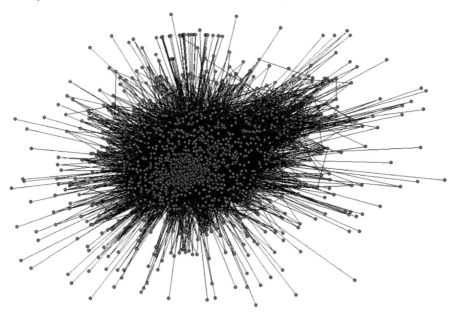

Figure 3. Visualization of relationships (black line) formed between forum members (red bullets) over the complete time span.

percent of all shortest paths between pairs of others. Specifically for carders, the mean betweenness is 0.3 percent (SD=0.013), whereas for those who do not operate as carders the mean betweenness is 0.05 percent (SD=0.001). This difference is found to be significant using a Mann-Whitney U-test (U=150923.5, p (one-sided) <.001). As we cannot account for interdependency of observations here, the significance of this effect might be overestimated. However, since the mean difference is relatively large and highly significant, we do consider this outcome as support for the hypothesis. Additionally, a separate analysis (results not presented) shows that carders are more likely to start relationships with members who offer data and with members who offer cash services on the forum's marketplace. This is in line with the expectation that carders are likely to be positioned in between data thieves and people who offer cash services as they need both kinds of resources.

To test the remaining hypotheses, we switch from analyses at the individual level to analyses at the dyadic level. The unit of analysis is the tie between a sender and receiver. The results for the logistic regression analysis are presented in Table 3.[1] Model 1 is used to test the hypothesized main effects. Since only 0.9 percent of the possible ties actually exist (see Table 1), effects on the probability of a tie will always be small, and as there are more than two million cases, effects are likely to be significant. Therefore, we will also look at the mean increase in probability of a tie with changes in the independent variables (i.e., marginal effects) to evaluate effects in terms of relevance. Note that these are the effects over and above the effect of the existence of a tie the other way around (reversed tie), as that has been controlled for.

The second hypothesis predicted that "carders have a higher outdegree in the network than other members—they send more messages." The significant positive coefficient of *carder* shows that the probability that a tie exists from a sender to a receiver increases if the sender is a carder. In other words, carders are more likely to have outgoing ties and thus a higher outdegree than others. The marginal effect reveals that if the sender is a carder, the mean probability to send out a tie is .0028 higher than for other members. Given that the mean probability of a tie is .0093, this increase is considered relevant and thus supports the second hypothesis.

The third hypothesis, "the more often a forum member offers services, the higher a forum member's indegree in the network, that is, the more messages a forum member receives," is also confirmed by the data. The significant positive coefficient of *services* shows that the more often a receiver offers services, the higher the probability of receiving a tie and thus the higher the member's indegree. For every time a service was offered by a receiver, the mean probability of receiving a tie increases by .0011. Considering the mean probability of a tie (.0093), this is a reasonable effect.

The fourth hypothesis, which predicted that "the better a forum member's reputation, the higher a forum member's indegree in the network, that is, the more messages a forum member receives," is also supported. The significant positive coefficient of *reputation* shows that the better a forum member's reputation, the higher the probability of receiving a tie and thus, the higher the member's indegree. For every positive rating (i.e., one additional unit on the reputation scale), the mean probability of receiving a

1. Unfortunately, we are not able to show to what extent the models can explain variation in individual centrality, as a suitable alternative for R^2 is not yet available for clustered logit models.

tie increases with .0004. Given that reputation scores vary from −2 to 40, this is considered a reasonable marginal effect.

The fifth hypothesis, "the higher a forum member's status, the higher a forum member's indegree in the network, that is, the more messages a forum member receives," is also supported. The significant positive coefficient of *status* shows that the higher a forum member's status, the higher the probability of receiving a tie and thus the higher the member's indegree. For every step on the ordinal status scale, the mean probability of receiving a tie increases by .0009. Given that the mean probability of a tie is .0093 and that status scores range from 0 to 10, this is considered a relevant marginal effect. Moreover, we know only the most recent status. Members actually might have had a very high status before their degradation to a negative status. As a consequence, some actors might end up with high indegree scores and a very low status, so that the presented status effect is likely to be underestimated.[2]

Conclusion

This study was undertaken in order to describe and explain differences in individual centrality within a cybercrime network using forum data. The results show that there is considerable variation in individual centrality. The network, however, does not show a typical core-periphery structure, with a small group of leaders connecting many otherwise unconnected peripheral members. The core is actually so extended that the center and periphery can hardly be distinguished. As expected, it appears that individuals can extend their own personal network quite easily in this virtual forum environment. Hence, many members are able to occupy a central network position. It has been argued that if central positions are not exclusive, the structural advantages disappear and central actors are not necessarily powerful [5]. The cybercrime network under study seems to be self-organized without central direction. In line with what law enforcement

2. If all cases with receivers with a negative status are excluded the marginal effects are indeed found to be much stronger. Further, the effect of services becomes insignificant. This is probably due to power reasons, as those who are excluded often are the ones who offered services.

agencies focused on cybercrime often claim, forum members primarily seem to operate on an individual basis and get involved in varying dyadic relationships [21, 22].

Although there is no clear core-periphery structure, considerable variation in individual centrality has been found. Hence, we set out to investigate determinants of central network positions using insights from online social networks, social capital, and criminology literature. The results show that carder activities, offered services, reputation, and status have the hypothesized positive primary effects on individual centrality. To be more specific, carders are more likely to actively contact many members, probably because they need to be involved in many ties to pursue their interests. In addition, carders are more often on the shortest path between two individuals than others, probably between data thieves and people who offer cash services. Forum members who offer services on the forum's marketplace are also found to be popular relations, especially those who offer web and security services such as spam and so called bulletproof hosting. Many others seem to be interested in their resources. Moreover, the results show that a better reputation and a higher status also make members more popular partners. It seems that reputation and status indeed can assist members in selecting partners.

In conclusion, it appears that forum members exchange resources in varying dyadic relations to make their own activities profitable. An online cybercrime forum seems to be comparable to a physical marketplace where people meet to buy and sell goods. Those who want to buy a lot have many relationships because they are very active on the market and those who offer interesting goods are popular contacts. Furthermore, being a reputable or high status actor also triggers relationships. In line with the "criminal enterprise model," the cybercriminals on the forum can be conceptualized as entrepreneurs who follow ordinary market rationality [10]. These findings explain why law enforcement agencies seem to struggle with finding hierarchies, groups, and *the* most important actors or leaders on this kind of forum [21, 22]. There are many well connected members, and restraining a couple of central actors by arrest does not immediately unravel the network and disrupt the system. Conceptualizing cybercrime networks as marketplaces could result in a more effective approach than applying investigative strategies that are common for physical networks in organized crime. This study indicates that the question should not be only *who* the central actors are, but especially *why* they are central. Understanding which characteristics make them central can help to touch on the vital parts of the cybercrime infrastructure. In the network under study, central actors who offer resources that are most exclusive and critical for the carding process can be important targets for law enforcement. It is interesting for future research as well as law enforcement agencies to examine this further and investigate which "stand" should be the first to be removed from the market.

As the phenomenon under study is relatively new, so is this field of research. This study contributed to the literature in several ways. First, in contrast to most empirical studies on cybercrime or physical criminal networks, this research is theory-driven. Bringing together scientific literature from the field of online social networks, social capital, and criminology as well as insights from law enforcement agencies dealing with cybercrime, results in a theoretical framework that has proved to be useful. Second, we used forum data, including direct measures of relations between cybercriminals as well as several individual characteristics, and proved the utility of this kind of data for social network analysis.

One of the limitations of the data is that only the most recent reputation and status are known. As a consequence, we were not able to test the hypotheses with dynamic data. This implies that we cannot make causal claims and do not know the actual direction of the relationships. It might be that members first need to have many relationships before they can get a better reputation and a higher status. Nevertheless, we have been able to evaluate network characteristics and empirically test relationships between individual attributes and network positions. However, as other cybercrime forums have hardly been studied, we do not know to what extent results can be generalized, and it would be interesting to examine similarities and differences between forums. Moreover, there might be overlap in members, and studying different kinds of cybercrime forums can provide a more complete and possibly more nuanced picture of the cybercrime underground economy. For example, peripheral members on this forum can be central members on another forum, or resources that are exclusive on this forum can be abundant on another forum. Furthermore, analyzing online communication activity outside forum environments could give more insight in how relationships develop after the first meetings on the market.

To conclude, this study shows that it is worthwhile to evaluate forum data from a social network perspective. The methods used can statistically confirm or reject existing ideas and lead to new insights as well. Further investigation would not only contribute to scientific literature, but a better understanding of cybercrime networks can also help to develop more effective strategies to combat this type of crime.

REFERENCES

[1] R. S. Burt, (1992). *Structural Holes*, Harvard University Press, Cambridge.
[2] R. S. Burt, (1997). "The contingent value of social capital," *Administrative Science Quarterly*, 42, pp. 339–365.
[3] R. S. Burt, (2000). "The network structure of social capital," *Research in Organizational Behaviour*, 22, pp. 345–423.
[4] V. Buskens and W. Raub, (2002). "Embedded trust: Control and learning," *Group Cohesion, Trust and Solidarity*, 19, pp. 167–202.
[5] V. Buskens and A. Van de Rijt, (2008). "Dynamics of networks if everyone strives for structural holes," *American Journal of Sociology*, 114, pp. 371–407.
[6] J. Coleman, (1990). *Foundations of Social Theory*, Belknap, Cambridge.
[7] K. S. Cook, C. Snijders, V. Buskens, and C. Cheshire, (2009). "Introduction," In K. S. Cook, C. Snijders, V. Buskens, and C. Cheshire, (eds), *eTrust: Forming Relationships in the Online World*, Russel Sage Foundation, New York, pp. 1–12.
[8] C. Dellarocas, (2003). "The digitization of worth-of-mouth: Promise and challenges of online feedback mechanisms," *Management Science*, 49, pp. 1407–1424.
[9] M. Fafchamp and F. Glubert, (2007). "The formation of risk sharing networks," *Journal of Development Economics*, 83, pp. 326–350.
[10] C. Fijnaut and L. Paoli, (2004). *Organised Crime in Europe: Concepts, Patterns and Control Policies in the European Union and Beyond*, Springer, Dordrecht.
[11] J. Franklin, V. Paxson, A. Perrig, and S. Savage, (2007). "An inquiry into the nature and cause of the wealth of internet miscreants," In *Proceedings of the 14th ACM Conference on Computer and Communications Security (CCS'07)*, Alexandria.
[12] C. Freeman, (1979). "Centrality in social networks: Conceptual clarification," *Social Networks*, 1, pp. 215–239.
[13] T. J. Holt and E. Lampke, (2009). "Exploring stolen data markets online: Products and market forces," Forthcoming in *Criminal Justice Studies*.
[14] H. Hu and X. Wang, (2009). "Evolution of a large online social network," *Physics Letters* A, 373, pp. 1105–1110.

[15] A. Josang, R. Ismail, and C. Boyd, (2007). "A survey of trust and reputation systems for online service provision," *Decision Support Systems*, 43, pp. 618–644.

[16] E. R. Kleemans, E. A. I. M. Van den Berg, and H. G. Van de Bunt, (1998). *Georganiseerde criminaliteit in Nederland: Rapportage op basis van de monitor georganiseerde criminaliteit*, WODC, Den Haag.

[17] P. Klerks, (2000). *Groot in de hasj: Theorie en praktijk van de georganiseerde criminaliteit*, Samson en Kluwer, Antwerpen.

[18] A. Kirman, S. Markose, S. Giansante, and P. Pin, (2007). "Marginal contribution, reciprocity and equity in segregated groups: Bounded rationality and self-organization in social networks," *Journal of Economic Dynamics and Control*, 31, pp. 2085–2107.

[19] E. R. Leukfeldt, M. M. L. Domenie, and W. Ph. Stol, (2010). *Verkenning cybercrime in Nederland 2009. Boom juridische uitgevers, Den Haag.*

[20] N. Lin, (1999). "Social networks and status attainment," *Annual Review of Sociology*, 25, pp. 467–487.

[21] NHTCU (2007). *Cybercrime, focus op high tech crime: Deelrapport criminaliteitsbeeld 2007*, Thieme MediaCenter, Rotterdam.

[22] NHTCU (2010). High tech crime: *Criminaliteitsbeeldanalyse 2009*, Thieme MediaCenter, Rotterdam.

[23] K. K. Peretti, (2008). "Data breaches: What the underground world of 'carding' reveals," *Santa Clara Computer and High Technology Law Journal*, 25, pp. 345–414.

[24] W. Raub and J. Weesie, (1990). "Reputation and efficiency in social interactions: An example of network effects," *American Journal of Sociology*, 96, pp. 626–654.

[25] W. Raudenbush and A. S. Bryk, (2002). *Hierarchical Linear Models: Applications and Data Analysis Methods* (2nd edition), Sage Publications, Thousand Oaks.

[26] H. G. Van de Bunt and E. R. Kleemans, (2007). *Georganiseerde criminaliteit in Nederland: Derde rapportage op basis van de monitor georganiseerde criminaliteit*, WODC, Den Haag.

[27] R. C. Van der Hulst and R. J. M. Neve, (2008). *High-tech crime, soorten criminaliteit en hun daders*, WODC, Den Haag.

[28] S. Wasserman and K. Faust, (1997). *Social network analysis: Methods and applications*, Cambridge University Press, Cambridge.

[29] T. Yamagishi, M. Matsuda, N. Yoshikai, H. Takahashi, and Y. Usui, (2009). "Solving the lemons problem of reputation," In K. S. Cook, C. Snijders, V. Buskens, and C. Cheshire, (eds), *eTrust: Forming Relationships in the Online World*, Russel Sage Foundation, New York, pp. 73–108.

Part III: Experiences

Securing IT Networks Incorporating Medical Devices: Risk Management and Compliance in Health Care Cyber Security

Nicholas J. Mankovich

The U.S. health care sector continues to grow, even in a difficult economic climate. In 2009, health care accounted for 17.3 percent of the gross domestic product [2]. The medical technology industry in the United States accounted for 6 percent of the total health care industry revenue in 2008 [22]. The U.S. Department of Homeland Security has identified the Healthcare and Public Health (HPH) sector as one of the thirteen critical cyber security infrastructures [18]. Cyber security threats can affect both the manufacturing and direct care delivery role of medical technology, as it contributes to the prevention, diagnosis, and treatment of disease.

Because of the increasingly digital nature of medical imaging, medical device output represents the highest volume of digital data in the medical record. Cost-containment efforts by governments and delivery organizations continue to improve through the ever-broader application and use of information technology. Much of this is achieved by the direct and rapid connection of health data sources to those caregivers who make health care decisions. Improvements are most easily seen in improved fidelity, timeliness, and utility of the data presented to medical decision makers, with a minimum of direct support staff time required for collection, transportation, and presentation. Often, this data flow requires large-scale interconnection of medical devices, hospital information systems, and IT networks within a health care system. All of this interconnection and drive to deliver "anywhere, anytime" exposes information technology systems to cyber security threats.

In practice, health care IT and the data therein is rarely an intentional direct target of a security exploit. More often, these systems and data become collateral damage in a broader cyber security attack seeking financial data or compromising basic IT assets for reuse in non-health security compromises (e.g., assemblage of bot armies for denial of service attacks or use in email spamming). However, the continuing growth in medical identity theft seems to indicate that health data security threats are rising as criminals discover how to monetize health information [20].

Nonetheless, there are many well-documented attacks significantly compromising health care organizations. In January of 2010, the Conficker worm led to the outage of approximately 10 percent of the health care IT infrastructure in Sweden [23]. A similar outbreak led to approximately 15 percent of New Zealand's health care system going offline in December of 2010 [3]. Prior large-scale outbreaks like the Code Red worm in 2001 or the Sasser worm in 2004 led to less-documented health care outages [13]. Although the events in Sweden and New Zealand were some of the most dramatic, small-scale outages happened across the United States and many other countries despite the sometimes more advanced security protections among institutions in those regions.

Medical Device Industry and Security

The rise of cyber security events over the past few years involving increasingly interconnected medical devices has led to actions within the medical device industry [15]. The Digital Imaging and Communication in Medicine standard (DICOM) [17] issued its first Security Profile in 2001 (PS3.1 Part 15). This standard specifies mechanisms used to implement security policies in the interchange between entities employing the DICOM standard.

A medical device without network connection or digital media exchange is generally highly secure. Security only becomes a serious concern when devices are interconnected, usually by attaching them to institution-wide networks. It is this creation of IT networks incorporating medical devices and health care information systems (i.e., the creation of medical IT networks) that poses a threat. Competent authorities responsible for safe and effective commerce in medical devices were also alerted to the risks of IT security in medical devices. In 2005, the U.S. Food and Drug Administration issued its "Guidance for Industry—Cybersecurity for Networked Medical Devices Containing Off-the-Shelf (OTS) Software," advising that it is the responsibility of medical device manufacturers to maintain cyber security maintenance plans in order to maintain safe and effective use of connected devices [16]. Many manufacturers have issued medical device security notices to their customers, often using the Internet to provide these documents (e.g., see Philips Healthcare "Product Security Policy Statement") [19].

Further, as the need for improvement in security has developed in the health care industry, the Healthcare Information and Management Systems Society (HIMSS) has formed work groups to address basic IT privacy and security, integrity of patient data, and medical device security. In response to the legal requirement that U.S. health care organizations engage in security risk management of electronic Protected Health Information (ePHI), HIMSS and the National Electrical Manufacturers Association (NEMA) created a template for documenting and communicating security attributes in the Manufacturer Disclosure Statement for Medical Device Security (called the "MDS2 form").

Relationship among Security, Privacy, and Trust

Before going on to review the current activities in medical device security and privacy, it is important to recognize and understand some subtle definitions and concepts. *Security* is a condition where controls are implemented to adequately protect data and system confidentiality, integrity, and availability. This is realized as a set of technical, administrative, or physical constraints (controls) on how systems or data are maintained. In the international community, we have tried to use the phrase "data and systems security" to promote understanding in translation to languages that do not linguistically distinguish the words "security" and "safety" (e.g., German *Sicherheit*, French *sécurité*).

Privacy is the expectation that information about the individual will be kept out of public view and in the control of the individual. If security is an attribute of systems and data control (often implemented as technical, physical, and administrative controls), privacy is the expectation of the individual about maintaining control of personal information. Security is thus a means to privacy control. Privacy rights in a society are often codified in law and regulation. A health care organization or a provider of medical devices, systems, or services must provide assurances of both security and privacy in order to build trust with its constituency and remain in compliance with law.

The concern for security and privacy in health care continues to lead to national and international initiatives designed to improve controls around sensitive data (i.e., a defined class of personal information that includes health data). The European Commission created Directive 95/46/EC to address the protection of personal data (European Parliament 23/11/1995). [5] In the United States, HIPAA (the Health Insurance Portability and Accountability Act of 1996) set out the Privacy Rule and the Security Rule for the control of all electronic Protected Health Information (ePHI) [21]. This has been further strengthened in subsequent regulations and legislation in, for example, the U.S. Health Information Technology for Economic and Clinical Health Act (HITECH). Privacy and breach notification requirements continue to grow as countries and states enact new protective legislation.

In the world of medical devices, my company, as a manufacturer, distinguishes *product security* as the management of products and services that support our customers in maintaining confidentiality, integrity, and availability of protected health information and the hardware/software systems that create and manage it. For the majority of our business, we act as a business-to-business supplier working for the Health Delivery Organization (HDO; hospital, clinic, or doctor's office), providing hardware, software, and services that support the customer's health care mission.

Security and privacy obligations reach a manufacturer's agenda due to law, regulation, contract, policy, ethics, or commercial pressure. This naturally leads the programs around privacy and security to be organized as

compliance programs. Of course, much of the legal and regulatory obligations bear directly on the HDO; the manufacturer's obligation is to serve
the requirement of the customer's compliance program. At Philips Healthcare, we organize our privacy and security programs around the elements
of compliance as elaborated in the U.S. 1991 Federal Sentencing Guidelines
for Organizations [1, 12]. From this guide created for judges by the U.S.
Sentencing Commission, an organization can see the six critical elements
that require active engagement and improvement to demonstrate an effective compliance program (see Table 1).

Table 1. Elements of a Diligent Security Compliance Program as Derived from the
U.S. 1991 Federal Sentencing Guidelines for Organizations.

	Requirement	Activity (evidence)
1.	Policies and procedures	Written guidelines for organizational and staff behavior (e.g., corporate policy for privacy and security)
2.	Training and awareness	Active training of staff and campaigns designed to encourage correct behavior (e.g., mandatory security course for all service engineers, mandatory privacy course for every employee at least every two years)
3.	Event management	A clear means to and demonstration of ownership of investigation and management of potential breaches (e.g., guidance for managing events and a clear documentation trail for privacy or security event investigation and resolution)
4.	Auditing and monitoring	Engagement of organizational auditing staff combined with ongoing monitoring of the state of compliance (e.g., periodic reporting to executive management on the status of the program; engagement of corporate audit on requirements of policy)
5.	Disciplinary standards for staff non-compliance	Policy around staff personal impact of non-compliance (e.g., evidence of having enforced such standards)
6.	Expert advisors	Clear appointment of advisors specified by policy or procedure (e.g., identified privacy and security experts; a program that improves their expertise and a demonstration of how the organization applies their expertise)

Why Is Security an Issue in Health Care Organizations?

In general, establishing state-of-the art security in health care organizations is a difficult task. Most health care organizations are highly focused on the delivery of health care services, and direct most risk management activities toward those life and death decisions made within the organization. Decisions are influenced by the cost-constrained nature of most nongovernmental health care organizations which usually operate not-for-profit enterprises. Cost reduction can push the organization to remove barriers to information flow. Security, on the other hand, usually means creating barriers. The art of security is accomplishing controls without removing essential information flow.

Medical device manufacturers experience a broad spectrum of demand for product security support. Although most requests for proposals received in my company contain security language of some sort, I have found that about 5 percent of health care delivery organizations are firm and are vocal about security. Another 20 percent demonstrate their concern for security in having appropriately detailed purchasing documentation but show flexibility. The vast remainder of health care organizations may have a few security requirements, but once provided with some security language in the response, they do not let deviations from their specifications hinder the purchase process. Purchasing decisions tend to be driven by caregivers and procurement/finance managers, not IT specialists. Most participants seem to assume that medical device and service providers take care of all security, but in reality there is little security dialog between the vendor and the health care organization. Because manufacturers pay much more attention to features that capture the imagination of the equipment or service decision makers, security functionality has, in many cases, been overlooked. This is especially striking in comparison to many other types of generic information technology equipment.

Gaps in security also stem from confusion about assumed roles and responsibilities of the provider organization and the manufacturer. Traditionally, decisions to purchase a device have been based on the equipment's ability to perform the specific health care task—usually with emphasis on the most up-to-date methods and the fastest, highest-quality results. In this drive to medical excellence, security often receives short shrift. Since the increase in network interconnectivity often comes from outside of the medical department actually purchasing the equipment (e.g., from the IT organization, medical records, billing), the purchaser often assumes that the IT organization will secure it, or more problematically, presume that the manufacturer somehow ensures the absolute security of the equipment. Additional pressures to connect hospital systems to the Internet too often open inadvertent attack paths directly into the diagnostic, storage, or therapeutic equipment.

The effect of combining this confusion with the slow, inexorable growth in device connectivity is that much of the installed medical equipment base is not ready to be connected in an open fashion on large-scale hospital networks. This leads to security solutions that put medical devices into protected network enclaves. This approach has been pioneered by the U.S. Veterans Administration with their 2004 creation and 2009 update of "Medical Device Isolation Architecture Guideline" [6]. This practical guide to the implementation of a protected isolation network can be found at the HIMSS Medical Device Security website [14].

A Risk-Management Framework for the Medical IT Network

As a result of a revision in the U.S. HIPAA Security Rule and its compliance deadline of April 2005, medical device manufacturers in 2004 started receiving a large variety of requests for security and privacy control information from health care organizations. HIMSS and NEMA organized a Product Security Work Group to address the need for standardized medical device information in order to properly assess security risk associated with the health information managed by those devices. In December of 2004, the work group issued the Manufacturer Disclosure Statement for Medical Device Security (MDS2) [8]. This three-page questionnaire with manufacturer notes provides basic information on a device's security management of electronic Protected Health Information (ePHI) and the administrative, physical, and technical safeguards associated with the respective medical device. This became the first-line communication instrument for manufacturers describing security attributes of equipment including their security/privacy controls.

However, security concerns were but one of the critical risk issues surfacing before medical device regulatory authorities. The failure to manage the complexity of the interaction of medical devices with other networked information technology was causing some real harm to patients. This was being seen as hospital-generated voluntary incident disclosures to regulatory authorities. Many potential sources of harm inherent in the interconnection of medical devices have been known for the past decade, but this point was graphically underscored in the Health Information Technology (HIT) testimony of Dr. Jeffrey Shuren, Director of the U.S. Food and Drug Administration [7]:

> Nevertheless, certain HIT vendors have voluntarily registered and listed their software devices with the FDA, and some have provided submissions for premarket review. Additionally, patients, clinicians, and user facilities have voluntarily reported HIT-related adverse events. In the past two years, we have received 260 reports of HIT-related malfunctions with the potential for patient harm—including 44 reported injuries and 6 reported deaths. Because these reports are purely voluntary, they may represent only the tip of the iceberg in terms of the HIT-related problems that exist.

Even as the MDS2 form became the industry standard for communicating security- and privacy-specific information, competent authorities around the world were becoming increasingly concerned about the risks inherent in using information technology to connect medical devices within a health care enterprise. In December of 2005, Brian Fitzgerald of the FDA called together a diverse group of regulatory staff, manufacturer representatives, hospital biomedical engineers, standards writers, and consultants. He used that meeting to summarize the documented incidents where harm was traced to improper interconnection of medical devices within a hospital IT network. Specifically, he asked the group, "How can we help the health care organizations properly manage risk associated with safety, effectiveness, security, and privacy?"

The FDA meeting and the recognized need for a broadly useful risk management framework led to the formation of Joint Working Group 7 [4] between the International Organization for Standards Technical Committee on Health Informatics (ISO TC215) and the International Electrotechnical Commission's Subcommittee on the Common Aspects of Electrical Equipment Used in Medical Practice (IEC 62A). This working group then began to create an international standard: "Application of risk management to IT-networks incorporating medical devices" (ISO/IEC 80001-1:2010) (International Organization for Standardization 2010) [9]. This standard represents a kind of "process transfer" building on the quality systems employed in manufacturing medical devices to recommend a set of good practices for managing risk throughout the entire life cycle of the connection of a medical device to a Health Delivery Organization's IT network infrastructure. Figure 1 depicts the life-cycle process elements of 80001-1 [10]. It is aligned both with medical device manufacturer risk process standards (e.g., ISO 14971, IEC 60601) and IT service management standards (e.g., ISO/IEC 20000).

As the title indicates, the 80001-1 standard is all about risk management. It is not about technical requirements for networks, interconnections, or interoperability. Instead, it is about establishing processes that bring the right people together to develop a way of working that balances the benefits of interconnectivity with the potential risks to safety, effectiveness, and data and systems security (including privacy). It provides the means for health care organizations to do their best in managing a new or ongoing medical device network connection through the careful consideration of these three key properties in the light of the connection's intended use. The 80001-1 standard provides a framework for managing these risks in planning, implementing, maintaining, changing, and decommissioning a medical device IT network connection through the full life cycle. Although the majority of risk management processes and decisions happen within the HDO, there are also roles and responsibilities for IT vendors and medical device manufacturers to support the HDO activities. The 80001-1 scope statement is [11]:

> Recognizing that MEDICAL DEVICES are incorporated into IT-
> NETWORKS to achieve desirable benefits (for example, INTEROPER-
> ABILITY), this international standard defines the roles, responsibilities and
> activities that are necessary for RISK MANAGEMENT of IT-NETWORKS
> incorporating MEDICAL DEVICES to address SAFETY, EFFECTIVE-
> NESS and DATA AND SYSTEM SECURITY (the KEY PROPERTIES).
> This international standard does not specify acceptable RISK levels.

So, for the first time, a common international framework exists, placing security and privacy on common ground and in the same discussion as safety and effectiveness. The Joint Working Group 7 remains active in producing

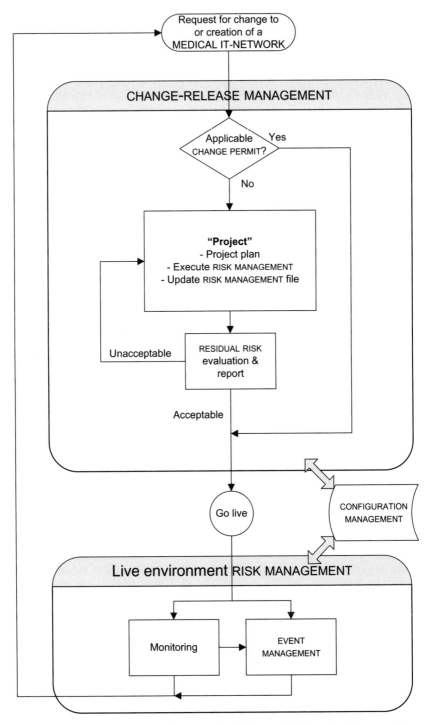

Figure 1. Overview of the life cycle of a medical IT network. Note the alignment of processes to ISO 20000. (Image reprinted by permission from the International Electrotechnical Commission [IEC] from its International Standard IEC 80001-1 ed.1.0 [2010].)

guidance documents in the form of IEC Technical Reports, as there remains a great deal to learn in the practical implementation of this standard. As of this writing, three proposed guidance documents are in circulation for feedback from the community of users:

> 62A/719/NP "Step by step risk management of medical IT-networks; practical applications and examples"
>
> 62A/720/NP "Guidance for the communication of medical device security needs, risks and controls"
>
> 62A/721/NP "Guidance for wireless networks"

Communicating and Understanding Security Risk

Within these proposed guidance documents is a framework for the communication of security risk, intended to provide a means to communicate the IT network security-related capabilities of a medical or IT device from the manufacturer to the organization responsible for connecting it to the HDO's network. It is a general description of the most relevant security attributes that the HDO might want to understand before starting its own security risk evaluation prior to medical device connection. The outcome of a security risk evaluation would be the planning of additional security mitigations, the acceptance of risk, or the rejection or disablement of certain features or functionality that might lead to risks not easily mitigated and thus deemed unacceptable.

Although the particular details of a risk management process will be determined by the scale and scope of the organization and its connection project, the kind of security capabilities that are treated in the guidance document is shown in Table 2. (See the final IEC 80001 Security Technical Report for the final standardized list.) It is presumed that the IT vendor or medical device manufacturer will produce a report of controls addressing each of the nineteen security areas that serves as the beginning of a dialog about risk management as the HDO plans its equipment procurement and subsequent interconnection project. This dialog may extend to the vendor/manufacturer providing on-site personnel support to the risk management team of the HDO as it executes its process.

Conclusion

Security of health care delivery is an important element of cyber security activities at both a national and institutional level. Medical devices comprise a mission-critical and highly valued part of the maintenance and restoration of health and well-being. Medical device security is an important component of the overall security stance of the health care sector.

Information technology has provided tremendous gains in productivity and effectiveness in health care. The understandable urge toward interconnection of medical devices and health care systems with IT networks has produced a great deal of complexity within health care organizations. This complexity, when not properly managed for risk, can cause harm to people. To manage the risks along with realizing the maximum benefit, health care industry leaders, including health care organizations, government authorities, medical device manufacturers, and others have created a framework for medical IT network risk management. The ISO/IEC 80001-1:2010

Table 2: A List of Medical Device Security Characteristics Used in Risk Managing the Connection of the Device to a Medical IT Network.

Abbreviation	Security characteristic
ALOG	Automatic logoff
AUDT	Audit controls
CNFG	Security feature configuration
DTBK	Data backup and disaster recovery
EDID	Health data de-identification
EMRG	Emergency access
ESTO	Health data confidentiality for storage
GUID	Security guides
INTG	Health data integrity and authenticity
MALP	Malware detection/protection
PEAU	Person and node authentication
PLOK	Physical locks on device
PRUP	Cyber security product upgrades
RDMP	Third party components in roadmaps
SAHD	System and application hardening
SERV	Service access security
TXCR	Transmission confidentiality
TXIN	Transmission integrity
UUID	Unique user ID

provides a risk management framework to deal with the complexity of risk to safety, effectiveness, and data and systems security.

There remains a great deal of work to be done to realize effective risk management in this highly interconnected IT environment. A framework is just the beginning. The remaining responsibility rests with health care providers and IT practitioners—they have to make it work for their patients and institutions.

The publication of the ISO/IEC 80001-1:2010 is just the release of a voluntary standard. If this becomes usable and worthwhile in securing data and systems, we can expect leading organizations to adopt it in practice. As this gets broader exposure, perhaps hospital accreditation organizations will work it into an accepted practice and even a required set of processes. This could, in some parts of the world, also be turned into harmonized standards or be adopted by reference into regulatory language. Much depends on the success of the standard in reducing risk at reasonable cost and on how the benefits are perceived by the public and politicians. For now, the future of risk management in medical device security has a clear path but remains open to implementation planning and cost-effective innovation.

REFERENCES

[1] Ethics Resource Center. (2005, December 31). "Federal Sentencing Guidelines," [Online]. Available: http://www.ethics.org/resource/federal-sentencing-guidelines

[2] D. Brown. (2004, February 4). "Health-care sector grew as economy contracted in 2009," [Online]. Available: http://www.washingtonpost.com/wp-dyn/content/article/2010/02/03/AR2010020303622.html

[3] "Computer virus cripples Waikato DHB." *New Zealand Harald* (2009, December 17). [Online]. Available: http://www.nzherald.co.nz/compute/news/article.cfm?c_id=1501832&objectid=10616074

[4] T. Cooper and S. Eagles. (2009, January 9). "IEC SC62 A Joint Working Group 7: Application of risk management to information technology (IT) networks incorporating medical devices," [Online]. Available: http://www.iecstandards.org/dyn/www/f?p=102:14:0::::FSP_ORG_ID:2471

[5] European Parliament. (1995, November 23). "European Parliament and Council Directive 95/46/EC of 24 October 1995 on the protection of individuals with regard to the processing of personal data and on the free movement of such data," *Official Journal of the European Union L 281*, pp. 31–50.

[6] H. Haislip, J. Deltognoarmanasco, T. Tepp, H. Stockley, and D. Pettit. *Medical Device Isolation Architecture Guide, v2.0.* Washington, D.C.: US Department of Veterans Affairs, August 2009.

[7] Health Information Technology (HIT) Policy Committee, Adoption/ Certification Workgroup. (2010, February 25). Jeffrey Shuren, Director of FDA's Center for Devices and Radiological Health. Testimony, Washington, D.C. [Online]. Available: http://healthit.hhs.gov/portal/server.pt/gateway/PTARGS_0_10741_910717_0_0_18/3Shuren_Testimony022510.pdf

[8] HIMSS Medical Device Security Work Group. (2004, December 17). "Manufacturer Disclosure Statement for Medical Device Security—MDS2 Version 1.0," *HIMSS Medical Device Security Manufacturer Disclosure Statement for Medical Device Security.* [Online]. Available: http://www.himss.org/content/files/MDS2FormInstructions.pdf

[9] International Organization for Standardization. *Application of risk management for IT-networks incorporating medical devices—Part 1: Roles, responsibilities and activities.* International Standard, Geneva, Switzerland: ISO, 2010.

[10] The author thanks the International Electrotechnical Commission (IEC) for permission to reproduce Information from its International Standard IEC 80001-1 ed.1.0 (2010). All such extracts are copyright of IEC, Geneva, Switzerland. All rights reserved. Further information on the IEC is available from www.iec.ch. IEC has no responsibility for the placement and context in which the extracts and contents are reproduced by the author, nor is IEC in any way responsible for the other content or accuracy therein.

[11] IEC 80001-1 ed.1.0 Copyright © 2010 IEC Geneva, Switzerland. www.iec.ch

[12] J. M. Kaplan. (2001 November/December). The sentencing guidelines: the first ten years. [Online]. Available: http://www.singerpubs.com/ethikos/html/guidelines10years.html

[13] G. Keizer. (2004, May 7). Sasser worm impacted businesses around the world. *Network Computing.* [Online]. Available: http://www.networkcomputing.com/data-protection/sasser-worm-impacted-businesses-around-the-world.php.

[14] Medical Device Security—Archive Reference Materials. (2010, December 2). [Online]. Available: http://www.himss.org/ASP/topics_FocusDynamic.asp?faid=101.

[15] E. Messmer. (2005, November 28). Hospitals' patch fears on the wane. *Networkworld Security.* [Online]. Available: http://www.networkworld.com/news/2005/112805-medical-security.html

[16] J. F. Murray. (2005, January 14). Guidance for Industry—Cybersecurity for Networked Medical Devices Containing Off-the-Shelf (OTS) Software. *Medical Devices—Device Advice: Device Regulation and Guidance.* [Online]. Available: http://www.fda.gov/MedicalDevices/DeviceRegulationandGuidance/GuidanceDocuments/ucm077812.htm

[17] National Electronic Manufacturers Association. (2010). Home Page. [Online]. Available: http://medical.nema.org/

[18] Office of the President of the United States. (2003, February). The National Strategy to Secure Cyberspace. *Department of Homeland Security.* The White House. [Online]. Available: http://www.dhs.gov/xlibrary/assets /National_Cyberspace_Strategy.pdf

[19] Philips Healthcare Product Security. (2010, December). Product Security. [Online]. Available: http://www.healthcare.philips.com/main/support /productsecurity/index.wpd

[20] Ponemon Institute, *Benchmark Study on Patient Privacy and Data Security.* Sponsored Report, Traverse City, MI: Ponemon Institute, 2010, November.

[21] "The Health Insurance Portability and Accountability Act of 1996," *US Public Law 104–191.* Washington, D.C.: 45 CFR Part 160 and Part 164, August 21, 1996.

[22] The Lewin Group, *State Economic Impact of the Medical Technology Industry.* Washington, D.C.: The Lewin Group (for AdvaMed), 2010.

[23] S. Wiinberg. (2010, January 19). e-virus ett ständigt hot mot patienten! (e-viruses a constant threat to the patient!). *Notes from the MTF seminar on e-viruses in Stockholm January 19, 2010.* [Online]. Available: http://www.mtf .nu/Utbildning/Reportage/e_virus/reportage_e_virus20100119.htm

Computer Forensics from a Law Enforcement Perspective

Kevin Kelly

Computerized information has become an integral part of our everyday lives as it has created a new perspective, almost a parallel virtual world that reflects our own physical world. This cyber world is abundant with evidence, especially when it comes to combating, investigating, and solving almost any crime, not just crimes committed with the assistance of computers, but all types of crimes. Computer forensics is the science of gathering and preserving electronic evidence or digital artifacts off electronic media. This digital evidence needs to be forensically preserved so it can be presented in litigation, to prove innocence, guilt, or to resolve a legal matter. The science of computer forensics involves anything where digital information, ultimately stored as ones and zeros, needs to be extracted from some type of media and preserved for use as evidence.

Evidence from a cyber crime scene can be easily duplicated, altered, and contaminated if not handled properly. Not every computer expert knows how to handle digital artifacts. Acquiring digital information forensically from a cyber crime scene requires a knowledgeable technician. The three fundamental areas a technician has to focus on to properly gather evidence are to first know exactly where, when, and what you are allowed to search and seize, provide thorough documentation, and finally to preserve the digital artifacts.

Information technology has become more accessible as new products catch up with advancements in technology such as the miniaturization of computerized components, quicker processing speed, easier-to-use interfaces, Internet connectivity, and increased storage capacity. Connecting a computerized device to the Internet gives you global access to a wide variety of resources, but also has the potential to allow you to be a cyber victim simply because criminals worldwide now have access to you.

Several new technologies have helped facilitate global communications. Improvements and commercialization of satellite technology, increased performance of personal computers, advances in networking technology, fiber optics, wireless communications, and broadband have let us communicate instantaneously almost anywhere in the world and at a moderate cost. This portability and ease of use in computer products, such as the iPhone and the additional functionality of cellular phones, have made it possible to imprint financial records, contacts, documents, accounts, and other electronically stored material on these devices. This information can yield a tremendous amount of evidence that can impart the user's personality (likes and dislikes), network of associates, spending habits, calling habits, global positioning, and much more.

The Internet is another dimension of the cyber world that is a complex entity and that poses many problems with jurisdictional and legal issues when it comes to prosecuting crimes. Cyber interconnectedness using the Internet has become a predominate force not only in personal usage, but in practically any public and private sector infrastructure. The economy, energy grids, financial industry, telecommunications, health, public services, water, and food and agriculture are all now incorporated into cyberspace. Any attack on these infrastructures will most likely include a cyber attack [4]. In order to protect Internet users and secure these infrastructures, new laws and treaties need to be implemented to help combat some of these crimes happening in cyberspace's global environment.

Incorporating Security into a Business Plan

There are two distinct approaches to the way evidence is gathered in either a public or a private investigation of a cyber event [3]. Both approaches have one thing in common: there is an *expectation of privacy* afforded persons against unnecessary and illegal searches and seizures [1]. The ability to search electronic media can be thought of as a right, or an expectation of privacy issue. Private investigations and public investigations use completely different methods. Private investigations are handled with civil or administrative processes, whereas public investigations are dealt with in a criminal arena. That is, generally speaking *private* equates to civil matters and *public* pertains to criminal matters. A private search relies on an organization's company policy and warning banners, whereas a public search is directed by the Fourth Amendment of the U.S. Constitution.

Successful private investigations depend on the employee's not having an expectation of privacy in the workplace. There are two protections that an organization needs to implement to ensure they have a right to view, protect, and analyze any electronic data on their computer systems. These protections are accomplished by implementing a strongly worded company policy, and by implementing a warning banner to lower an employee's or contractor's expectation of privacy on the organization's systems. The policy should state that the organization's computers are for the organization's use, and users should expect no privacy using the systems. This policy must be included in the employee manual and made available to all employees in addition to being incorporated into agreements with outside contractors. A warning banner should also be displayed to every user, every time the user logs on to or gains access to a company's computerized systems. A typical warning banner states: "This computer/network is for authorized users only. There is no expectation of privacy to users of this system. The company reserves the right to monitor, retrieve, and disclose any information stored or transmitted over this system." When the company policy is spelled out and a warning banner is implemented, a private investigation and examination of electronic media can be performed anytime by company officials. This is the most important point to be made in this chapter. There are many companies that do not consider this factor until it is too late.

Investigations of government employees fall into the same category as private investigations when it comes to use of the government's computer systems. To conduct a proper investigation and mitigate any civil liabilities, employees must not have any expectation to privacy. Proper policies must be implemented, and warning banners must appear when employees boot up or log in to government computer systems.

Public investigations are conducted as investigations of criminal activity and are handled by government agencies. The subject of an investigation in the United States is protected by the Fourth Amendment to the U.S. Constitution, which defends an individual against unlawful searches and seizures. A public investigator first must have probable cause that a crime has been committed, and must define the area to be searched and the items sought after. In addition, the investigator must show that all investigative steps have been exhausted and must focus on specific crimes, suspects, data, time periods, and the likelihood the evidence still exists. After these thresholds are met, the public investigator swears before a judge that the information provided is true and is issued a search warrant. The public investigator now has the right to search and seize digital evidence as defined by the search warrant signed by a judge.

There are exceptions to the protections provided by the Fourth Amendment. For example, if a person is dead, he has no expectation of privacy and his computer may be searched without a search warrant. Also, if consent is given, and parents can give consent to search their minor child's computer, or when evidence is found in plain view (such as contraband on the screen including child pornography or other incriminating evidence), no warrant is required to seize the computer, but a search warrant may be required to continue looking for evidence. Even though these exemptions are valid reasons to search without a warrant, the best policy is to secure a search warrant whenever possible in a criminal matter.

Note that if a private investigation becomes criminal and law enforcement is called in to investigate the crime, then the more stringent guidelines of public investigations take precedence and the Fourth Amendment rules to search and seizure apply. A criminal case will have to be developed and a search warrant drawn up if necessary.

It is an important step in acquiring digital evidence to have policies and warning banners in place prior to a cyber incident, and to show probable cause that a crime has been committed to obtain a search warrant. These steps will ensure that you have a right to seize and search through the evidence as soon as practical.

Utilizing Policy and Government Resources

Not too long ago a typical criminal investigation started by gathering witnesses, interviewing them, processing a crime scene using traditional forensic methods, building leads by finding associates and witnesses, interviewing them, developing motives (why) and modus operandi (how), picking up a suspect, and confronting him with this information to try to get a confession. Today, one of the first things to do is locate and examine any digital media such as closed circuit TV cameras, cell phones, laptops, computers, and so on in addition to performing traditional investigative techniques.

Preserving this information correctly and securely can make or break the final outcome of a case. This is where computer forensics comes in. Computer forensics uses many specialized tools to preserve evidence, find deleted information, crack passwords, locate fragments of a file, and automate searches. A qualified technician should perform these functions.

Digital evidence can help us determine whether a crime or offense has been committed, and how a crime has been committed. It can develop an investigative lead, verify a witness statement, identify suspects, link offender-victim relationships, and find abuse of company assets. But handled incorrectly, it can yield more questions than answers.

A good case to review is that of Pittsburgh Steelers quarterback Ben Roethlisberger's sexual assault case in Milledgeville, Georgia. The nightclub where the incident took place had video footage of what transpired in the club the night of the incident (CBSnews.com, March 26, 2010). The police viewed the footage and left the crime scene without taking the electronic evidence from the digital closed circuit cameras. Several days later, the police came back to retrieve the video footage, but it was taped over and lost forever. This footage had evidence that may have corroborated some of the victim's or subject's story. Even though a physical forensic examination of the scene was performed and evidence was secured, the investigators neglected to take into account that digital evidence at the crime scene can be transient.

Where, when, and what you can search and seize is a major part of the investigation. Investigators and prosecutors looking to tighten up their cases have asked to have a "computer forensics" done for anything that dealt with a computer. However, there are legal mechanisms in place that tell you what you can and cannot do in searching for digital evidence. For instance, getting hit over the head with a keyboard does not give you the right to search, and gather evidence off, a computer. A search warrant might not give you permission to search offsite storage facilities not mentioned in the warrant. So preparation becomes an important part of computer forensic investigations. Routers, proxy servers, and switches may save logs to remote servers. Knowing every place your evidence is located is instrumental in conducting a proper and thorough investigation.

In another case, I was involved in a raid on an escort service (prostitution ring) that had three locations, and there was a search warrant obtained for each location. We had the option to search each location one at a time, or to spread our resources and hit all three locations at the same time. We correctly decided to hit all three locations simultaneously. Each location had video feed of the other two locations; the computers at each location were on the same network and had the capability to delete information at the two other locations. So, had we executed the search warrants separately there was the strong possibility of losing digital information at the other sites and possibly at the original site as well. Just as with other investigations, digital investigations need tactical planning to ensure safety and success. This is particularly important now when most organized criminal enterprises that are making money use the latest digital technology.

Investigating crimes committed with the assistance of a computerized device also requires knowledgeable investigators. Typical computer crimes include network intrusions, corporate and computer espionage, identity theft, theft of intellectual or artistic property, computer fraud, child pornography, and child endangerment. Not only do cyber crime fighters have to know how to investigate the specific computer crime, they also have to know the technology used to commit these crimes, from how information is stored on a hard drive to tracking down IP addresses over the Internet. All these factors have to be mastered in order to pursue criminal activity in cyberspace.

Protection from Cyber Adversities

Anyone who exercises some degree of access, control, or responsibility for managing information systems must now also possess some fundamental skills for securing those systems. Network administrators, application developers, and even end users need to incorporate security measures into their mindset. It is not only the government's responsibility to protect public resources, especially cyberspace resources; private enterprises and end users must also take the necessary precautions to secure their systems.

Detecting cyber crimes and securing cyberspace are important to keeping our cyber environment safe. The government must take a leading and active role in hunting down and capturing criminals who use cyberspace as a vehicle to perpetrate crimes. A government must know the depth and severity of crimes being committed so it can develop the tools needed to combat the crimes.

There are many dilemmas to be faced in securing information at all levels—from users to organizations to the government agencies. The Wikileaks scandal brings several of these issues to the forefront. In the Wikileaks scandal, stolen classified information from government and corporate organizations was posted on the Wikileaks website for the public to view. The first issue is how to prevent the theft of classified information by an organization or government employee, and protecting information from the inside. The next issue involves legally prosecuting the website owner for posting classified information. Is the government employee a hero for uncovering corruption, or a criminal for disclosing classified information? Has a law been violated? Are there jurisdictional issues involved in enforcing a law? Is the website posting a crime, and is it extraditable? Are the necessary tools in place to gather evidence, prosecute the perpetrator, and stop the criminal acts? In a global society, with respect to what is legal and what is not, the answers to these questions vary greatly from crime to crime and in different jurisdictions.

Advancement in technologies, especially in computer-related crime, wireless communications, and the Internet has made many of our laws outdated, obsolete, and unenforceable for combating high technology crimes and preserving the global security of a nation. There is no simple clear-cut solution, as many of these issues have been around prior to the technological advancements, although they are more prevalent now with the interconnectedness everyone is afforded over the Internet.

Both corporate and government computer system administrators and IT professionals need to have a rudimentary understanding of cyber statues that are applicable to their line of work at the local, state, and federal levels. A good reference to the laws of the United States, several other countries, and to international law is available at http://wings.buffalo.edu /law/bclc/resource.htm. Each site linked to from this site is maintained by the respective governmental organization.

Regional law enforcement in the United States uses an informal network of cyber crime enforcement groups to share information and collaborate on cyber investigations, which usually span many different jurisdictions. Groups such as the cyber cop portal (https://cybercop.esportals.com/), High Technology Crime Investigator Association (www.HTCIA.org), the FBI sponsored InfraGard organization (http://www.infragard.net/) as well as many others all foster fraternal bonds and help to close out the investigations of the many participants. So many federal and international crimes are committed that there are often not enough law enforcement officials in the affected jurisdictions to handle them. For example, the FBI may not pursue a fraud crime unless it meets the threshold of $50,000. Crimes not meeting specific thresholds are handled by regional or local law enforcement agencies that have to overcome jurisdictional obstacles by using these fraternal organizations.

The approach to jurisdictional cooperation modeled by the Department of Homeland Security is an example of a step in the right direction, but more coordination is needed at the local level. The approach to homeland security is based on the principles of shared responsibility and partnership with Congress, state and local governments, the private sector, and the American people, as well as a centralized control of federal agencies. Coordination of local and state resources allows for proper communication of pertinent issues and important information. In the information technology age, a centralized system that is structured for immediate use would be beneficial, but in the United States the task would be colossal at best. On a global level it is even more difficult; nevertheless coordination on a global level must happen to secure cyberspace.

Over the last three decades computers have evolved and have become more of a threat by automating several of the tools hackers use. All a potential cybercriminal has to do is open a search engine such as Google, search for "vulnerability" and the particular operating system he is looking to exploit, and within an hour or so he could be hacking into a computer on the other side of the globe. Information is power, and the Internet affords anyone the ability to seek that power not only for productive purposes but destructive ones as well. The number of attacks is so large, and the sophistication so great, that it is hard to determine how to best defend your systems.

One of the greatest challenges facing the law enforcement and IT communities is handling the multi-jurisdictional issues involved in high tech crimes. After a computer system is connected to the Internet, the computer system and everything attached to it becomes globalized. System administrators need to be aware of this because their system may be attacked from anywhere in the world, and not only from hackers, but from perpetrators of corporate espionage, nation-states, and organized crime as well. Attacks on industries and especially government agencies are high priorities of these subversive groups.

Computer professionals should be aware of what is going on in the world. Destabilization of countries produces some of the best hackers and social engineers in the world. The collapse of government often leads to localized corruption and proliferation of organized crime and black markets in these countries. Intellectuals who were on top of their food chain now have to find a way to earn a living in this environment and more and more are turning to the Internet and hacking into systems to gather information that can be traded or converted into big money. They look for anything of value on networks, such as personal information useful for identity theft, and proprietary, intellectual, or military information for sale over the Internet.

Subversive groups, terrorist organizations, or state-sponsored groups will deploy whatever resources are available for the sole purpose of destroying our country and way of life. Identity theft and online fraud have allowed criminals to fund terrorist activities. For more information, see the FBI testimony before the Senate Judiciary Committee, Subcommittee on Technology, July 2002 [5].

Countries affiliated with anti-American terrorist organizations view stealing from American companies as a way of redistributing the wealth unjustly taken from them. There is very little difference between terrorist groups and organized crime: both have exploited virtually every financing method at all levels, with identity theft at the root of most of these schemes. Most financial institutions had originally ignored the problem and looked at it as a cost of doing business. The latest figures estimate that identity thieves made off with approximately $47.6 billion in 2003 and $15.6 billion in 2006 [2]. Significant portions of the proceeds were having a direct or indirect influence on the major operations of these subversive groups. Americans are not the only targets; every country on the Internet is also a target.

Criminal entrepreneurs have discovered the lucrative business of stealing identity, which robbed victims of an average of approximately $5,000 in 2003 and $2,000 in 2006 [2]. These people are a potential group, in that they are not unified but they exist separately, unorganized as a whole but are out there hammering away at our institutions for financial gain through the Internet. Often information can be bartered for over the Internet, allowing criminals to set up black markets where they can collaborate. They help set up highly sophisticated and elaborate scams aimed at our financial institutions. In addition, several smaller scams are used to involve innocent actors, such as having people looking for work by answering employment ads for jobs to work at home by accepting deliveries, or picking up cash at Western Union and sending the proceeds overseas while keeping a percentage for themselves. Not only are these scams hard to trace to find the culprits, but when you do find them, they're often in countries where the local government is so corrupt that little can be done, and there is little support for solving crimes that help the local economy.

There has to be a unified effort by businesses, private citizens, and governments to become educated about securing their piece of cyberspace. Setting up policies to allow cyber investigations to be conducted over your corporate assets and securing any computer environment is a basic starting point. Government should always be vigilant in securing cyberspace by testing systems, gathering intelligence, providing laws and treaties to protect cyberspace, and developing liaisons with foreign governments to protect each other from global threats.

The IT sector has to be given the appropriate resources to preserve its cyberspace. IT administrators have to have the right personnel and the right tools such as hardware, software, and administrative policies to secure their computer systems. Government systems by nature have a tremendous wealth of information that can be exploited in several ways. After a system is in place it must be constantly improved and upgraded. Computer processors, memory, and hard drives have been doubling their speed and capacity every year or two since 1980, with connectivity speed also advancing at an accelerated rate. Automation of programming languages has also improved dramatically over the last thirty years. These factors create an environment conducive to allowing hackers to create and deploy attacks against computer systems. As hackers improve their success in gaining access to systems, IT professionals must remain vigilant about procuring the resources necessary to secure the information entrusted to them.

Some systems are not secured, and even if they do not have valuable information on them, they can be used for creating a distributed network for password cracking, to direct a denial of service attack against our cyber infrastructures, or for several other scenarios, all of which can be prevented. In these situations where the cyber environment can be used as a weapon, the government may have to step in and require systems attached to the Internet practice due diligence and due care to provide a minimum amount of security.

Conclusion

The cyber world, with its abundance of information, is a new venue to locate valuable evidence needed to fight crime. These digital artifacts need to be appropriately acquired by knowledgeable computer forensic technicians to properly preserve the integrity and validity of the evidence. Investigators, whether working on a public or private investigation, are relying more and more on the information found in cyberspace. Not properly preserving digital evidence can have a devastating outcome on investigations. A qualified computer forensic technician is needed to properly secure digital artifacts from computer systems. If you have to extract information from a computer system without a forensic technician, document everything you do.

Make sure warning banners and organizational policy explicitly detail the rights a user can expect when using a company's computer system. The removal of a user's expectation of privacy on a company's digital assets will allow for proper investigation and monitoring of these systems.

Enforcing computer laws poses many problems with legal and jurisdiction issues. The Internet is global in scope and the real world has many boundaries. Each of these boundaries has different jurisdictions, laws, traditions, and enforcement on local, state, federal, and international levels. Cooperation on all levels must be achieved in order to successfully enforce laws and prosecute crimes committed over the Internet. International cooperation is needed to create laws and treaties for cyber crimes happening in the global environment know as cyberspace.

REFERENCES

[1] E. Casey, *Digital Evidence and Computer Crime, Second Edition*, Elsevier Academic Press, 2004.
[2] *Federal Trade Commission—2006 Identity Theft Survey Report*, November 2007.
[3] B. Nelson, A. Phillips, and C. Steuart, *Guide to Computer Forensics and Investigations*, Course Technology Cengage Learning, 2010.
[4] SANS, Cyber Warrior Manual, 2004.
[5] Terrorist Financial Review Group, FBI before the Senate Judiciary Committee, Subcommittee on Technology, July 2002, [Online]. Available: http://www.fbi.gov/news/testimony/the-fbis-support-of-the-identity-theft-penalty-enhancement-act

Computer Crime Incidents and Responses in the Private Sector

Edward M. Stroz

Companies can fall victim to various types of computer crime and accidental incidents. One type of incident is a data breach in which information a
company has in its possession is stolen or in some way improperly released.
An obvious example we have all read about is the theft of individual social
security account numbers (SSANs) or credit card numbers. There are other
forms of data breach, too, such as the theft of trade secrets.

If we just focus on data breaches, we see that the size of this problem alone
is striking. According to figures posted in 2010 by www.privacyrights.org,
494,647,283 records were breached from 1,651 separate instances of data
breach made public since 2005 [1]. An often-cited example of a large data
breach was announced by Heartland Payment Systems in 2009. Press reports
indicated that approximately 130 million records may have been compromised [2]. Many other companies, governments, and not-for-profit organizations (such as TJX, Choicepoint, and the Veterans Administration) have
reported data breaches in which millions of records were exposed or stolen.

Sometimes the cause of the breach is a lost laptop, the theft of a computer hard drive or backup tape, or the action of a disgruntled employee. In other cases, data breaches are brought about by malicious computer software (malware) that has been installed surreptitiously into a company's computer network. You may have read press reports about what have become known as advanced persistent threat (APT) technologies, which have been an eye-opener for business executives. While your company might appear to be running just fine, and even profitably, your company's data may have been targeted and compromised using hidden techniques. And one of the more disconcerting properties about information is that, unlike money or physical property, it can be stolen without depriving the owner of that which has been stolen. The theft happens by copying rather than removal.

Who Does This, and Why?

Those who seek to steal information are sometimes referred to as *threat agents*. Threat agents can have different motivations for what they do. Some intruders are motivated by financial profit, and they may be interested in stealing personally identifiable information (PII) to sell to identity thieves, or intellectual property such as trade secrets or research and development information (industrial espionage). Sometimes an intruder tries to sell the information back to the victim in an extortion scheme.

Other intruders may be motivated by ideology. Types of attacks from this type of intruder can include state-sponsored attacks, and are sometimes committed by trusted insiders of the targeted company. An insider attack can also be the work of a disgruntled employee, as appears to be the case of the WikiLeaks incidents involving the theft of diplomatic cable communications. Having some degree of understanding about the motivation of the attacker is important in developing an effective response to the incident, and in preventing these problems in the future.

Important Response Actions and Considerations

If the incident under investigation is believed to be from an attacker external to the victim company, it is important to start by identifying the computers that are likely to have been compromised by the attack. The process for identifying the compromised computers will have to be tailored to the specific computer network involved, and often depends heavily on a dialogue with trusted members of the IT staff, and on a good data map. (More information about the data map is in the "Useful Tips for Preventing Incidents and Effectively Responding in the Event They Happen" section later in this chapter.) From there, a forensic copy (or forensic image) of the data from the compromised computers is usually made to preserve the data that would otherwise be altered or lost.

The forensically preserved data are then analyzed for the presence of malware and data artifacts that could indicate whether the malware was activated and, if so, to what degree. For example, logs are analyzed and key events from those logs are correlated in an attempt to establish how, when, and where data were touched, copied, and transmitted. Other data that are not recorded in logs also have to be taken into consideration. This analysis requires someone with training and expertise in this type of specialized investigation. The information stored on the compromised computers is then assessed, usually with the involvement of legal counsel, to determine how sensitive it is and whether it triggers a legal reporting obligation. If malware is found to be present on a company computer, that malware has to be removed properly (but will remain in the data preserved on the forensic image as evidence that could be used later in court). This work typically takes days or weeks and is used in attempts to determine the nature, extent, and timing of the data breach, or to establish evidence that a breach did not occur or was limited.

If the incident under investigation is believed to be from an attacker internal to the victim company, the goals are largely the same (i.e., to determine the nature, extent, and timing of the data breach). The computers that were accessed by the attacker, or by the technique employed by the attacker, should be forensically preserved. The analysis steps will be tailored to the facts, as always, and can be used to gather evidence used in applying for a temporary restraining order covering the insider.

In instances of insider abuse, it is also important to take into consideration physical evidence that is usually not relevant when compared to an attack by an outsider. For example, if keycards are used to control access to physical office space, the records generated by such a keycard system should be preserved for later analysis. The same considerations should be given to surveillance camera recordings. In some instances that we have seen, it is important to check for physical signs of forced entry into rooms or computers. It is important that these investigative steps take into account the culture of the company and the disruption it can cause to business operations. It can also be helpful to keep in mind that a thorough investigation will not only competently gather evidence of any wrongdoing that might have occurred, but can also be useful in excluding certain employees from further suspicion.

In instances of lost or stolen computers, it is important to try to recover the computer itself and, if successful, examine the computer in an effort to try to determine what information, if any, was accessed while it was not in the possession of its rightful owner. For this reason, the analysis of the computer should be done using a forensically-sound approach that will not alter data and will produce evidence that can be admissible in court. Also, the search actions used with the data should take into account the various ways information can appear in computerized records (e.g., social security account numbers and credit card numbers may or may not use dashes or spaces to separate fields).

Useful Tips for Preventing Incidents and Effectively Responding in the Event They Happen

It is generally not realistic to expect a technological silver bullet, or to train employees so they will be eternally loyal and unfailingly vigilant. What is more realistic is to ensure your IT department knows where your company stores its information. A reasonably accurate and up-to-date data map for your company will show what type of information is stored where, and can help managerial discussions about policies and procedures over such data. In this regard, corporate policies, procedures, and personnel training should be current and reviewed by management for effectiveness in their design. It can be helpful to have a computer penetration and vulnerability testing program, perhaps carried out by an independent party. Corporate policies should include an incident response plan. Computer network operations should have robust, secure, and monitored logging, with the monitoring network logs stored in a centralized location in searchable form. Security and incident training should also be part of the policies and procedures.

A solid incident response plan will include the following elements:

An outline of responses planned for each type of incident.

Assigned duties and emergency contact lists.

For incidents that are believed to be perpetrated by an insider, a lock-out plan to cover the suspect.

A plan for documenting procedures and findings, and for gathering and storing evidence under a suitable chain-of-custody.

Liaison with legal counsel and human resources professionals.

Other considerations include having the response plan checked periodically to keep it fresh and ensure it is consistent with other corporate policies and procedures, including those related to privacy. It is also very important to have an implementation policy for preservation orders governing electronic and other forms of evidence. Many clients also find it useful to have a public relations plan that can be activated in the event of an incident.

Conclusion

The following is a list of ten observations that I have found to be helpful in providing advice to our clients:

1. Attacks that appear to originate from foreign countries may actually be using that country as a relay point. An experienced investigator should keep an open mind and work from facts. Don't fall into the trap of demonizing other countries in your investigation.

2. Victims of computer crimes are increasingly tempted to find fault with their service provider companies, especially those that are cloud-based, for not offering two factor authentication for logging-into the applications being offered. Offering username and password access may not be considered sufficient protection by your customers.

3. Companies are reluctant, for valid security reasons, to allow outsiders to come into their offices and then connect to their internal networks. Companies increasingly are expecting visitors to provide

their own access or are setting up separate networks just to accommodate visitors. It is safer to have only authorized and patched
computers on a network.

4. Government is increasingly becoming the source of intrusion
   detection notification and alerts to victimized companies. The "tap
   on the shoulder" is the way you may learn of a problem and you
   should have a plan in place for responding responsibly.

5. The investigative approach that is selected in a given incident
   should start by asking whether the incident is still active, because the
   answer to that question will have a significant impact on the correct
   response. An active incident is more complex to manage than an
   incident that is not active, such as a lost laptop.

6. An independent expert firm hired to help a victimized company
   will be more effective if that firm operates with the attitude that
   they are advocates for gathering complete and accurate evidence,
   and its impartial interpretation. Investigative theories should be fit
   to facts, not the other way around. The advocacy on behalf of the
   victimized company is more effectively carried out by their
   lawyers.

7. Putting your data in the cloud doesn't mean that you should forget
   about where they are located. The data you place in the cloud are
   stored in computers somewhere, and where they are stored will
   invoke legal considerations associated with that venue.

8. If the incident involved the installation of malware, that malware
   may need to be reverse engineered so that the code used in an attack
   can be investigated to determine its potential and the peculiar type
   of evidence it might leave behind when it is deployed or activated.

9. If a given event doesn't appear in a security log, it does not necessarily mean that the event definitely did not happen. Sometimes
   malware is designed in such a way that it will bypass logging or clean
   up after itself.

10. A given computer, such as your laptop or iPhone, is usually filled
    with data from different sources and having differing property
    rights. Such intermingled data may require that they be carefully
    organized into different category types and by the handling, protec-
    tion, and limits on who can view them. That reality is making it

important to have investigative actions taken using a protocol that will be thorough while recognizing and respecting the various types of information present.

REFERENCES

[1] Privacy Rights Clearinghouse, www.privacyrights.org
[2] Alleged International Hacker Indicted for Massive Attack on U.S. Retail and Banking Networks, (2009, August 17). Department of Justice. [Online]. Available: http://www.justice.gov/opa/pr/2009/August/09-crm-810.html

Information Technology for a
Safe and Secure Society in Japan:
Toward a Cyber-Physical Solution

Kazuo Takaragi

This chapter introduces the topic of information technology for a safe and secure society in Japan, showing an emerging trend toward a cyber-physical solution. Notable security incidents in the United States and Japan, and the Japanese national strategy for information security, are discussed first, followed by a discussion of emerging security technologies in Japan (such as a multiple risk communicator), cryptography, and other core elements.

Introduction

Japan, like every industrialized nation, has experienced a great rise in notable security incidents. To deal with this increasing problem, the National Information Security Center has been established; its goal is to create a National Cyber Security Strategy for Japan. Japan's overall strategy is to be one of the world's most advanced countries in terms of cyber security by 2020.

Notable security incidents: Notable security incidents have occurred every six or seven years since 1982 (Figure 1). In 1988, Internet worms caused an estimated loss of about $111 million over several days in the United States. In 2007, personal information leakage caused a total loss of about $22.2 billion in one year in Japan [17]. Figure 1 shows that the economic loss due to security incidents is increasing geometrically. Cases in Japan include physical forgery coupled with unsecure information handling.

Recent security incidents in Japan: Several illegal immigration cases in which fake fingerprints were used have occurred in Japan since January 2008. At least eight people have arrived in Japan from an Asian country and used fake fingerprints to evade the biometric checks at immigration control [19]. Furthermore, at a university in Tokyo, a flash memory device containing information about more than 10,000 students was taken by an academic staff member to be used at home. Information in the files on the device was leaked onto peer-to-peer networks through a personal computer used by the staff member at home [18].

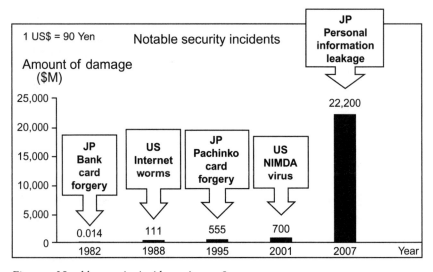

Figure 1. Notable security incidents since 1982.

Security market in Japan: The security market in Japan is still growing despite the global economic downturn. For example, the world market for security devices and tools was $3.728 billion in 2009 and is projected to be $5.305 billion in 2013 (Figure 2) [12]. The global market share of Japan was about 13 percent in 2007 (Figure 3) [13].

National Information Security Center: The National Information Security Center (NISC) was established in Japan in April 2005. Under NISC, the Information Security Policy Council (ISPC) was formed in May 2005. It makes decisions on the national strategy for information security for governmental agencies, critical infrastructures, businesses, and individuals. It also cooperates with foreign countries to establish an international reliance for security [7].

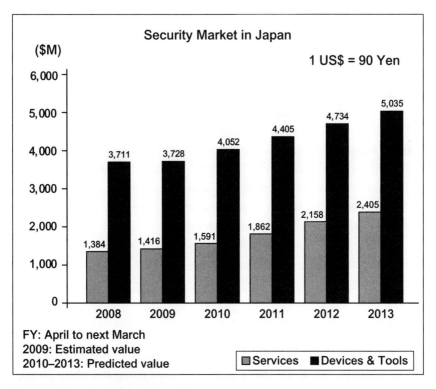

Figure 2. Security market in Japan [3].

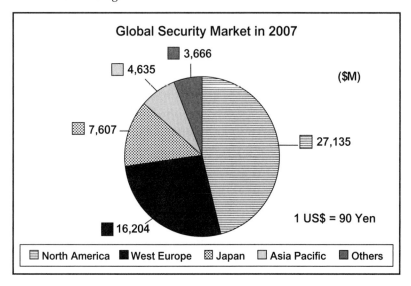

Figure 3. Global security market in 2007 [4].

National Cyber Security Strategy in Japan: A comprehensive national security strategy for Japan for 2010 to 2013 was published by the NISC in May 2010. It depicts the current problem of cyber attacks and their countermeasures. In June 2010, the NISC published a draft of the plan "Secure Japan 2010." Its goal is for Japan to become one of the world's most advanced countries in terms of information security by 2020 [7]. IT risk management by stakeholders is considered important. The plan strives to achieve security measures acceptable to those stakeholders. The plan describes government plans to facilitate smooth risk communication among stakeholders, such as concerned organizations and government agencies in charge of critical infrastructures.

Emerging Security Technologies in Japan

This section covers various emerging security technologies in Japan, beginning with an overview of security technology in a situation with two levels. Then, a system controller called a multiple risk communicator is described.

Technologies such as advanced cryptography, finger vein authentication, and RFID are also covered. Various issues related to mitigating information leakage are then discussed.

OVERVIEW OF SECURITY TECHNOLOGY

There are two levels of security technology. One is the component level, such as finger vein authentication and RFID. This level is explained later in this section. The other level is the system and management level. An example of an emerging technology at this level is a multiple risk communicator. A multiple risk communicator is useful for obtaining a consensus among many stakeholders in setting up a large system when multiple kinds of risks are present. The multiple risk communicator [14], devised by Professor Ryoichi Sasaki of Tokyo Denki University, who is also with the NISC in Japan, can be explained as follows. Security technology encompasses a broad spectrum, from the physical, such as finger vein authentication, to the cyber, such as information leakage damage reduction technology. Each of these elements has a risk measure. The multiple risk communicator meets as a group to discuss risk measure and determine appropriate action.

There is a precondition to be considered in executing the multiple risk communicator: trust should first exist among the stakeholders. Robert F. Hurley described a trust relationship between an employee and his or her boss [15], shown in Table 1. If the risk tolerance is low for the counterpart, then more time should be spent explaining options and risks and so on. In this example the counterpart is a person, but in general it does not necessarily have to be. It can be a machine, animal, or even something not of this earth.

For example, consider a scenario in which there is conflict between Country X and a different country, Country Y. Country X wants a rare metal located in Country Y. However, Country Y refuses to allow Country X to exploit the rare metal because doing so will damage the sacred tree standing at the site of the metal. In the next phase of the scenario, Country X wants to solve their problem at the expense of Country Y. Then in the next phase, neither makes mutual concessions—instead they fight. As a result, Country X is obliged to retreat. Country Y is successful in defending itself. However, the final situation and the damage caused may not be satisfactory for either party.

Table 1: Creating a Trust Relationship between Employees and Employers

If this factor is low	Then you should
Risk tolerance	Spend more time explaining options and risks Offer some sort of safety net
Security	Find ways to temper the risk inherent in the situation Expect to invest time in raising comfort levels
Alignment of interests	Focus on the overarching strategy, vision, and goals Shape a culture that reinforces doing the right thing for the enterprise
Capability	Find ways to demonstrate competence in carrying out the task at hand Acknowledge areas of incompetence and compensate by sharing or delegating responsibility

The original purpose of the risk communicator is to search for the optimal measure while considering related risks, costs, and benefits, and to help stakeholders reach an agreement. When there are two or more risks, especially conflicting ones, the multiple risk communicator is used to solve the problem.

OVERVIEW OF THE MULTIPLE RISK COMMUNICATOR

The multiple risk communicator developed by Professor Sasaki with the support of Hitachi and other companies in Japan consists of fault tree analysis and event tree analysis, among others. These analyses have often been applied in the safety evaluation of large-scale systems such as nuclear power plants. Using these analyses, the involved parties negotiate for policy agreement. Event tree analysis calculates the probability of each event sequence following an initial event. For example, if a successful Action 1 and a failed Action 2 occurred after the initial event, the second event sequence occurs with the occurrence probability of the initiating event multiplied by the success probability of Action 1 and the failure probability of Action 2.

A software tool for the multiple risk communicator was developed to assist specialists in doing an event tree analysis of a target system. Variations of the target system are sent to the tool so that various countermeasures can also be analyzed. An optimized result is displayed to help the involved parties obtain a consensus about a countermeasure (Figure 4).

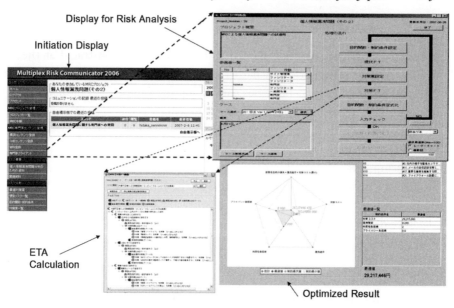

Figure 4. MRC display for risk analysis.

The tool was applied to a real problem in Tokyo: the conventional design of the staff room in a junior school was thought certainly to lead to information leakage. However, an overly strong countermeasure may lead to another problem: The traditional ease of communication between students and teachers in the staff room may be lost if control of entry into the staff room is too strict. This will decrease the communication between students and teachers, which may make it difficult for teachers to be made aware of bullying and other incidents. To solve such a multi-risk-related problem, three groups of people were gathered from the government office, the board of education, and the junior school. The multiple risk communicator was applied in such a way that they discussed privacy versus other safety considerations on the screen shown by the tool. Finally, they obtained a consensus to take action to:

1. ensure sufficient reduction of the probability of personal information leakage, and
2. maintain teacher-student communication by employing loose entrance control to the staff room.

ADVANCED CRYPTOGRAPHY

Hitachi has a hash function named Luffa that was demonstrated at the NIST SHA-3 competition. A hash function is a cryptographic function that cryptographically compresses data of any size into a short number sequence. It is an essential function for achieving various security mechanisms. Luffa [2] features a sponge structure that has a large space for randomizing processing, which is advantageous especially during parallel processing. The evaluated hardware performance is now at the top of the fourteen candidates in the second round of the SHA-3 competition. With very fast hardware speed and moderate software speed, it achieves a good balance of performance and safety.

FINGER VEIN AUTHENTICATION

Finger veins are biometric information belonging to each individual. They do not change throughout a lifetime, and cannot be viewed by the naked eye [3]. When an infrared light is applied to a finger, the light is absorbed mainly by the veins in the finger. A shadow is reflected to a camera to enable a picture of the finger vein to be taken. Hitachi has developed a finger vein authentication product with good performance. In particular, the accuracy rate is very high: The chance of false authentication is about a hundred times smaller than that when using fingerprint technology [10].

Many kinds of private information, including such biometric information, should not be processed in a remote server computer in a network because of privacy concerns. A breakthrough for solving this problem was made by an IBM researcher, Craig Gentry. He developed a fully homomorphic encryption scheme that can execute any mathematical operations on encrypted data [1]. The privacy of the plaintext is preserved during processing.

Consider this: Two finger vein pictures are represented by pictures of a tiger and lion, respectively, and sent to a server. The server then calculates them in a homomorphic way. The result is a combination of the two: a *Liger*. During the process, the server cannot guess what the original finger vein pictures are. On the one hand, the client can decrypt the Liger to obtain the qualitative difference between the two original pictures. However, a huge amount of computing power is currently required for perfect operation, and may require more than one trillion times the usual encryption computing. To address this, Hitachi has focused on specific applications that use limited mathematical operations so that the scheme works in practice. Such an application is a typical example of biometric authentication.

A service system for privacy protection during the biometric verification process is now available in Japan. Pattern matching for finger vein authentication can be done at high speed on a normal server without privacy infringement [9].

SMALLEST RFID IN THE WORLD

The μ-chip manufactured by Hitachi is currently the smallest RFID in the world. It is 0.4 mm in every direction and has been applied to the wiring management system of a nuclear power plant in Japan. A more advanced μ-chip is now in a trial production process [11]. It is 0.075 mm in every direction. It exhibits pressure tolerance of up to 400 kg per square cm. This is especially useful for paper applications.

MITIGATING INFORMATION LEAKAGE

Another emerging technology in Japan is a technique for mitigating information leakage. It uses crawling technology to connect to peer-to-peer (P2P) nodes one after another. Winny is a notorious P2P application in Japan. Many information leakage accidents occurred after Winny was infected by a computer virus. After detecting the leaked information, our method using crawling technology prevents the spread of the leaked information by manipulating the system at the data level (Figure 5).

The method for mitigating information leakage was tested in a simulator of the Internet named StarBED that is located in the districts along the Japan Sea [8]. In the test bed, about 200,000 users ran Winny at the same time. When a leaked file is discovered, a key is generated to delete the file. About 100,000 dummy keys were sent in about 1,000 seconds in the simulation, making the leaked files harmless. The simulation experiment was conducted with the assistance of Hitachi Information Systems, Ltd. [15, 16]. Note that if the technique is applied to a real-life network, it would be necessary to sufficiently consider the effects on real society, including legal and ethical aspects.

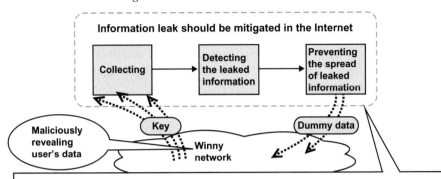

> **Step 1: Auditing by the use of crawling technology**
> ✓ Connecting to the P2P nodes one after another, followed by collecting the key data that points the file location on the Winny network
> **Step 2: Detecting the leaked information**
> ✓ Focusing attention on the characteristic of leaked file such as "including specified word"
> ✓ Detecting the key of related leaked file from a large amount of the keys obtained by the crawler
> ✓ Specifying the leaked file source and the leaked file distribution target
> **Step 3: Preventing the spreading of the leaked information**
> ✓ Replacing contents of leaked file with harmless data while the file name is retained
> ✓ Rewriting the part of the key (IP address or port number) to prevent downloding the leaked file
> ✓ Sending the falsified file and key (the dummy data) out over the Winny network

Figure 5. Mitigating information leakage.

Conclusion

There are several current standardization activities relevant to cyber security technologies and standards:

> **Cyber terrorism countermeasures:** Draft NIST IR 7628 Smart
> Grid Cyber Security Strategy and Requirements [6]
> **Hash functions:** NIST, "Second Round Candidates," NIST
> Cryptographic Hash Algorithm Competition, July 21, 2009 [5]
> **Biometrics:** ISO/IEC FDIS 24745 Biometric Information Protection

However, standards for multiple risk communicators have yet to be developed. In the examples given, there were basically two stakeholders: Country X and Country Y. The updated version of the multiple risk communicators will have the capability to handle thousands of stakeholders. It will use Ustream for audio-video communication and promoting understanding among the stakeholders, and Twitter for gathering opinions from stakeholders to find the most preferred solution to problems concerning various factors related to cost, risk, and benefit. Additionally, future multiple risk communicators will be able to supply stakeholders with customizable views on opinion visualization and allow data to be mined by opinion classification [17].

Acknowledgment

The author would like to thank Professor Sasaki of Tokyo Denki University for his helpful discussion about the usage of multiple risk communicators in a real society. The author also wishes to thank Mr. Yoshinobu Tanigawa, Mr. Kouichi Tanimoto of Systems Development Laboratory, Hitachi Ltd., and those involved with this study for their suggestions and assistance.

REFERENCES

[1] C. Gentry, "Fully homomorphic encryption using ideal lattices," *Symposium on the Theory of Computing (STOC)*, 2009, pp. 169–178.
[2] D. Watanabe, Y. Hatano, T. Yamada, and T. Kaneko, "Higher Order Differential Attack on Step-Reduced Variants of Luffa vi," *Fast Software Encryption*, FSE 2010, Springer-Verlag, LNCS 6147, 2010, pp. 270–285.
[3] Hitachi, Ltd. (2006). Finger Vein Authentication: White Paper. [Online]. Available: http://www.hitachi.co.jp/products/it/veinid/global/introduction/pdf/finger_vein_authentication_white_paper.pdf
[4] The National Institute of Standards and Technology, Computer Security Division, United States Department of Commerce. [Online]. Available: http://csrc.nist.gov/groups/ST/hash/sha-3/Round2/submissions_rnd2.html
[5] The National Institute of Standards and Technology, Computer Security Division, United States Department of Commerce. [Online]. Available: http://csrc.nist.gov/groups/ST/hash/sha-3/Round2/submissions_rnd2.html
[6] The National Institute of Standards and Technology, Computer Security Division, United States Department of Commerce. [Online]. Available: http://csrc.nist.gov/publications/nistir/ir7628/nistir-7628_vol3.pdf

[7] National Information Security Center. [Online]. Available: http://www.nisc .go.jp/eng/index.html

[8] The StarBED Project, Hokuniku Reserch Center. [Online]. Available: http://www.starbed.org/

[9] K. Takahashi and S. Hirata, "Cancelable biometrics with provable security and its application to fingerprint verification," *IEICE Transactions*, Vol. 94-A, No. 1, 2011, pp. 233–244.

[10] M. Terada, "Feasibility Study of DoS attack with P2P System," FIRST Symposium, Riga, Latvia, Jan. 19—Jan. 21, 2009.

[11] M. Usami, H. Tanabe, A. Sato, I. Sakama, Y. Maki, T. Iwamatsu, T. Ipposhi, Y. Inoue, "A 0.05 x0.05mm2 RFID Chip with Easily Scaled-Down ID-Memory," *ISSCC Digest of Technical Papers*, Feb. 2007, pp. 482–483.

[12] *Network Security Business 2009—Comprehensive Research Report*, Vol. 1 (Market Edition), Fuji Chimera Research Institute, Inc.

[13] NPO Japan Network Security Association (JNSA), "Information Security Market Survey."

[14] R. Sasaki et al., "Development and applications of a Multiple Risk Communicator," *Transactions of the Wessex Institute*. [Online]. Available: http://library.witpress.com/pages/PaperInfo.asp?PaperID=18773

[15] R. F. Hurley, "The Decision to Trust," *Harvard Business Review*, September 2006.

[16] R. Sasaki, N. Sugimoto, H. Yajima, H. Masuda, H. Yoshiura, M. Samejima, M. Funabashi, "Proposal for a Social-MRC: Social Consensus Formation Support System Concerning IT Risk Countermeasures," *Proceedings of IMS2010*, Korea, 2010, pp. 60–66.

[17] Survey Report of Information Security Incident 2007 (2009, March 31). NPO Japan Network Security Association. [Online]. Available: http://www .jnsa.org/result/2007/pol/incident/2007incidentsurvey_e_v1.0.pdf

[18] Daily Yomiuri [Online]. Available: http://yomiuri.co.jp.dy April 28, 2010

[19] Daily Yomiuri [Online]. Available: http://yomiuri.co.jp.dy, May 2, 2010

Part IV: Partnership, Policy, and Sustainability

Public-Private Partnerships Changing the World

Kathleen L. Kiernan and Dyann Bradbury

In 1961, United States President John F. Kennedy captivated the imagination of our nation when he announced before a joint session of Congress that "this nation should commit itself to achieving the goal, before the decade is out, of landing a man on the moon and returning him safely to the earth" [1]. NASA had yet to send a man into orbit to travel around the earth, and many were doubtful that even this could be done. In a clever way, he gave everyone permission to think differently. He gave everyone permission to imagine what we as a nation could accomplish.

Creativity is born from imagination. When we are given the opportunity to imagine, creativity flows freely, unfettered initially by the noise that accompanies tradition and by past organizational practice and biases; it's from the imagination that new ideas are introduced.

So, let's imagine for a moment that we can change the world. Imagine a world in which ordinary individuals from the private sector, law enforcement, government, and academia come together to share information openly about current threats, whether in the physical or cyber realm or at the hands of a violent non-state actor committed to violent terrorist activity. These ordinary individuals, often with extraordinary responsibilities, would come together with a sense of both a common and a shared mission, built on a foundation of trust and credibility. The mission's space would allow for the continuous exchange of information directly relevant to vulnerabilities, risk, mitigation, and remediation that is both sector-specific and transcends multiple sectors of the critical infrastructure.

Imagine if, while safeguarding legal and privacy requirements, law enforcement officials could harness the expertise of the private sector for assistance not just in early warning of anomalous activity but additionally for assistance in the response to and recovery from incidents in which elements of the critical infrastructure have been severely disrupted by man-made or natural causes. The usual process is initially to exclude all entry by non-law enforcement individuals, when in fact once the immediacy of the event is understood, the requirement is for the restoration of vital services, communications, and transportation—expertise resident in the private sector and perhaps, heretofore, invisible to the first responder communities. InfraGard as an organization has changed that equation and, as a result, is changing the world.

At the outset, InfraGard was developed to provide an opportunity for owners and operators of Critical Infrastructure and Key Resources (CI/KR) to share information about attacks, risks, vulnerabilities, mitigation, and remediation without any notional sense of seeking individual or organizational credit. The idea was predicated on the premise that threats could impact competitive businesses with equal ferociousness. The sharing of common threat information that could mitigate the threat while not jeopardizing competitive advantage was an idea that required a great deal of imagination mixed with a fair amount of trepidation—and it worked.

Since 1996, InfraGard has expanded to eighty-six chapters across the United States. Each chapter is geographically linked with an FBI Field Office and provides all stakeholders immediate access to experts from law enforcement, industry, academic institutions, and other federal, state, and local government agencies. By utilizing the talents and expertise of the InfraGard network, information is shared to mitigate threats to our nation's critical infrastructures and key resources in a timely, efficient, and effective way.

The fabric of world change resides within individuals who choose to contribute to a sounder and safer America. Paraphrasing anthropologist Margaret Mead, one should never doubt that a small group of thoughtful committed citizens can change the world. In fact, she said, it is the only thing that ever has! We continue to form partnerships and trusted relations with our local law enforcement, state, and local government, academia, emergency responders, and Department of Homeland Security (DHS) Protective Security Advisors (PSAs), ensuring that the voice of practitioners resonates at the policy level.

Organizationally, the charters for the FBI and DHS are based on the mission of protecting our nation's CI/KR. As the lead law enforcement agency for preventing terrorist attacks (including those targeting CI/KR), the FBI focuses on detecting, deterring, assessing, and countering threats and attacks. It also investigates and supports prosecution that may follow. DHS in the meantime seeks to provide a safe, secure, and resilient national infrastructure based on, and sustained through, strong public and private partnerships. When taken together, these complementary roles of the FBI and DHS encompass the entire risk reduction spectrum against terrorist acts to the nation's CI/KR. Collaboration and communication are the keys to protection. Providing timely and accurate information to those responsible for safeguarding our critical infrastructures, even at a local level, is paramount in the fight to protect the United States and its resources.

As an existing and structured public-private partnership, the InfraGard program consists of members that are owners and operators of our nation's CI/KR. An InfraGard member is a private-sector volunteer with an inherent concern for national security. Driven to protect their own industry and further motivated to share their professional and personal knowledge to safeguard the country, InfraGard members connect to a national network of subject matter experts (SMEs), communicate with federal law enforcement and government agencies through their local InfraGard chapters, and contribute to the security and protection of our national infrastructure from threats and attacks. InfraGard's strength and effectiveness are based on the subject matter expertise of its trusted membership. InfraGard members have access to the organization's secure internal portal powered by CyberCop, which enables the secure communication between members, local chapters, and local, state, and federal government agencies.

Critical Infrastructure and Key Resources Sectors

The National Infrastructure Protection Plan (NIPP) identifies risk as a combination of threat, vulnerability, and consequence. The activities across this entire spectrum must be well coordinated and effectively executed across government and with the nation's CI/KR owners and operators in order to avoid or mitigate the catastrophic consequences due to a failure of CI/KR facilities or systems. Critical Infrastructure and Key Resources are vulnerable to cyber and physical attack. What we can do is help to identify the risks, threats, and vulnerabilities and begin to mitigate and remediate, each of us within our own businesses and homes. We can help to educate the public about both cyber and physical security. We can indeed help to improve the security posture of our nation if we each do our part. We have much to give back to our country. It is our duty as citizens to make our country safer. These assets, for which we are responsible, are the foundation upon which our country depends for survival. It is time we work together to address these issues. It all begins through partnerships, information sharing, and hard work. Organizations dedicated to these missions are only the first step—it takes the selfless committed citizen to make it work.

I believe that our problems are both technical and ethical. We need to stop and do what is right. Do we rush a piece of code out the door, forsaking secure coding practices because we are up against a deadline? Does profitability trump the safety and security of our country? It shouldn't, but so often it does.

Each of you has it in you to make a difference. We are in this together, and only if we are united can we meet these challenges head on. Only if we are united will we begin to change the world.

Conclusion

Changing the world for the better is often thought impossible—there is always so much noise with day-to-day living and, given a grim economy, day-to-day survival. Yet we call on busy people to take on another important responsibility for homeland and national security where it matters most—in the far reaches of the homeland, away from the nation's capital—

and we ask for volunteers! We are constantly humbled by the selfless dedication of our members across the country who choose to make a difference without complaint. And in the process of giving back, they do change the world.

REFERENCES

[1] J.F. Kennedy, (1961, May 25). Historic Speeches—John F. Kennedy Presidential Library & Museum. *Address to Joint Session of Congress May 25, 1961.* (2011, March 10) [Online]. Available: http://www.jfklibrary.org/JFK/Historic -Speeches.aspx

Cyber Security: Protecting Our Cyber Citizens

Preet Bharara

We all recognize the importance of using every law enforcement tool at our disposal to combat threats to our cyber security. Computer networks—including the biggest one of all, the Internet—are crucial infrastructures supporting our global economy.

We in the Southern District of New York are critically aware of this. You need look no further than Wall Street, the nation's vital financial industry, for an obvious illustration of this fact. The securities exchanges, traders, and brokers rely heavily—if not exclusively—on computer systems to make billions of dollars worth of transactions each day.

Long gone are the days of paper-strewn trading floors and ticker tape (although as a Yankees fan, I hope there's always enough ticker tape around for the Yankees annual World Series parade down Broadway). Instead, we have automated trading systems residing on computer servers in Manhattan and elsewhere, linked by fiber-optic cable to traders around the world who place orders by computer.

While the Internet can often seem like a hazy concept, in truth it's an enormously complicated physical web of servers, databases, wires, and routers that reaches into nearly every home, business, and government agency in this country. It provides all of us with instantaneous communication with anyone in the world, nearly limitless applications, and unparalleled convenience and efficiency. And just as the Internet's flexibility and reach have radically transformed the way we communicate and conduct business, its pervasive nature has opened up new avenues for criminals to victimize us—even in our own homes.

The Cyber Crime Threat

The Internet is simultaneously full of promise and fraught with peril.

As I've learned over my last year as the U.S. attorney, the threats are all too real. These threats range from individual phishing emails that trick victims into disclosing personal and financial information, to computer viruses and keystroke loggers that steal victims' passwords, to online bank accounts, to the large-scale theft of account information from financial institutions, to denial of service attacks, to the theft of valuable intellectual property over the Internet.

Given what we have seen lately, the laptop and the modem are rapidly replacing the handgun and the ski mask as the preferred tools for robbing a bank.

A cyber attack could have an enormous, possibly devastating, impact on our economy. One need only consider a successful denial-of-service attack on the computer systems used by any of the stock exchanges on Wall Street to get a sense of what would happen.

Cyber crime also has the potential to cripple individual businesses and services. Consider, for example, what happened in January 2010 to Baidu, Inc. (www.baidu.com), the biggest Internet search engine in China and the third largest in the world. Hackers using the name "Iranian Cyber Army" took over Baidu's domain name and redirected users to a web page declaring that the site had been hacked. Baidu has alleged that the interruption to its operations cost it millions of dollars in lost revenue and damaged its business reputation. And Baidu's experience is not unique. A month earlier,

in December 2009, the popular micro-blogging site Twitter (https://twitter
.com) suffered the same attack, preventing millions of Twitter users from
accessing their accounts.

Although access to both Baidu and Twitter was restored in a relatively
short time, these examples demonstrate just *one* way that cyber crime can
instantly harm millions of users.

In monetary terms, cyber crime's impact is no less dramatic. A 2007 report by the Government Accountability Office estimated the direct economic harm from cyber crime to be on the order of $67 billion, a figure we
can all agree has only grown larger since then.

As the amount of money involved in cyber crime has grown, so too has
the sophistication and organization of cyber criminals. We see them becoming specialized, honing their skills in a particular area of expertise. By spe-
cializing, cyber criminals can develop a reputation, establish steady client
relationships with multiple criminal partners, focus their time and energy on
improving their goods and services, and increase their profit potential.

These specializations include coders, who create the malicious programs
that are necessary to steal valuable data like bank account information;
vendors, who trade and sell the stolen data; cashers, or money mules, often
unsophisticated individuals who use the information to illegally withdraw
cash from compromised accounts; and money launderers, who often hide
the criminal origin of illicit proceeds by converting them into digital
currency.

We also see today's professional cyber criminals using their own underground social networking sites to communicate, organize, and scheme together. Just like the old-world social clubs used by La Cosa Nostra—the
traditional Mafia, in the cases I used to prosecute as a new assistant U.S.
attorney in the Southern District—these hidden Internet forums are exclusive and by invitation only. New members must be vouched for by existing members. But once on the inside, cyber criminals can easily make new
contacts, buy, sell or trade contraband, and keep up-to-date on the latest
crime techniques.

To be sure, there are some obvious yet crucial differences between traditional organized crime and the new breed of cyber criminals. Just as the
Internet has increased the efficiency of communication and commerce, it
has also allowed cyber criminals anywhere in the world to work together
easily without respect for jurisdictions or national boundaries, rather than

in a more restricted geographic area like New York City's notorious Mafia families. And while Mafia members and cyber criminals each get to choose their own nicknames—although you're not liable to find too many "Fat Tonys" or "Little Larrys" in cyberspace—the Internet allows cyber criminals to operate with something of which old-school mobsters can only dream: near-complete anonymity behind their instant-messaging account number or nickname.

Cyber criminals don't need to resort to the old-fashioned walk and talk.

So it goes almost without saying that the investigation and prosecution of cyber crime represent the absolute cutting edge of criminal law enforcement. Given the unique challenges presented by this type of crime, I am glad that we have the best minds in government and industry working on this problem.

Later in this volume, FBI Director Robert Mueller writes on safeguarding our cyberspace. There is no one in America who better understands— and has more urgently emphasized—the need for cyber security: to protect our citizens, our economy, and our homeland.

In New York, we're lucky to be able to work with the FBI, which is at the absolute vanguard of the emerging fight against cyber crime. The FBI's New York Field Office is the largest in the country, and, recognizing the critical threat to us all from computer crime, it has greatly increased the resources devoted to the problem. Most impressively, the FBI in New York has created an entire cyber crime branch, headed by some of the most experienced agents around. Within the branch are specialized units to address cyber terrorism, computer intrusions, and fraud committed over the Internet. My office works closely with our colleagues at the FBI to address the cyber crime threat. But we can't do it alone. Coordination between law enforcement and industry is often the key to solving cyber crime cases.

What We Are Doing to Fight Cyber Crime

Before discussing how and why we should work together, I want to give you a sense of some of the work that the assistant U.S. attorneys in my office— the federal prosecutors in the Southern District of New York—have been doing to combat cyber crime. The Southern District places a high priority on prosecuting cyber criminals.

As a result, the Complex Frauds Unit has been created. Its mission is to handle the prosecution of cyber crimes and other sophisticated fraud schemes. Assigned to that unit are a number of prosecutors who have received special training in the nuances and difficulties of cyber crime investigations and prosecutions. But not just these assistant U.S. attorneys work on cyber crime—all of the twenty prosecutors in the Complex Frauds Unit make it a priority to build cyber crime cases. And it's not just the Complex Frauds Unit that is handling these matters—cyber crime cases are being worked on across the office—in the organized crime unit, the terrorism unit, and even by junior prosecutors. So as we shine a spotlight on these offenses, my office is committed to expanding our cyber crime prosecutions.

A significant part of my office's cyber crime strategy has been to develop and strengthen our relationships with foreign prosecutors and law enforcement agencies to target cyber criminals wherever they are found and to put them on notice that they are at risk of prosecution, no matter how safe they think they are. Cyber criminals work effortlessly across borders; that means we have to do the same.

As I've mentioned, the Internet's unique ability to connect people globally has made cyber crime one of the most international of criminal activities. And all too often, while international borders are obstacles to effective law enforcement, criminals see these boundaries as opportunities to confound prosecution. My office recognizes that close coordination among international law enforcement partners is crucial to overcoming the transnational challenge posed by cyber crime and to eliminate any safe havens for cyber criminals.

As a result, over the past two years the Southern District of New York has worked on investigations with a number of foreign partners, including Estonia, Romania, the Ukraine, Belarus, Lithuania, the Netherlands, the UK, the Czech Republic, and the Dominican Republic. Many of these investigations are ongoing.

Significant Cases

Working with our law enforcement partners in the United States and abroad, we have had a string of successes in the war on cyber crime. Let me mention a few to give you a sense of where we are, where we are going, and what is possible.

A recent example of this office's international pursuit of cyber criminals is the prosecution of Dmitry Naskovets, a Belarusian. Naskovets, together with a fellow national, Sergey Semashko, operated CallService.biz as an illegal service for online identity thieves. As you know, many financial institutions and Internet-based businesses have in place security measures to verify the identity of customers who wish to make transfers or withdrawals from bank accounts or make purchases with credit cards over the Internet. Among other things, customers may be required to confirm their identity to service representatives over the telephone. To beat this security measure, Naskovets and Semashko would provide the services of English- and German-speaking individuals to online identity thieves. The English and German speakers, using stolen account and biographic information supplied by the identity thieves, would pose as authorized account holders and confirm fraudulent transactions over the telephone.

According to advertisements for CallService.biz on underground Internet forums used by identity thieves, the CallService.biz website had performed over 5,400 confirmation calls to banks. In April of 2010, the Southern District of New York and the FBI, together with the Czech Republic, Belarus, and Lithuania, conducted a coordinated international takedown of CallService.biz's illegal business. Czech police arrested Naskovets in Prague, where he was living. Simultaneously, law enforcement in Belarus arrested Semashko. And Lithuanian law enforcement seized the computers on which the CallService.biz website was hosted. At the same time, the FBI in New York seized the CallService.biz domain name. Rather than just take down the domain name, however, the FBI put up a banner announcing the seizure.

I can tell you that the seizure banner, which is what online identity thieves who think they are visiting the CallService.biz website now see, has caused considerable anxiety in the online criminal world, enhancing the deterrent effect of the arrests of Naskovets and Semashko.

Indeed, we have since used domain name seizures and banners to similar effect with another law enforcement partner, Immigration and Customs Enforcement, in an operation called "In Our Sites," which targeted websites that distributed illegal copies of movies, music, games, software, and television shows over the Internet. On June 30, 2010, we seized seven domain names—with names such as TVShack.net and ThePirateCity.org that don't exactly hide what these sites are doing—and replaced them with seizure banners. In the weeks that followed, several other websites that distributed infringing contents have unilaterally stopped doing so. We believe that was in part because of the tremendous amount of attention that the banners attracted online. That is another heartening example of the power of deterrence in this area.

Here's another example: this office is also prosecuting Ilya Boruch, who ran a company called Bidding Expert in Queens, New York. On his website, Boruch claimed to offer various financial services to his customers, including converting cash into WebMoney, so-called digital currency that can be sent anonymously over the Internet.

Boruch is alleged to have taken more than a million dollars in cash from two individuals who were part of a conspiracy to fraudulently withdraw money from ATMs in New York City using stolen account information that they received from another criminal over the Internet. Boruch would meet the two individuals on the street in Queens and Brooklyn, where they would literally hand him bags of money. After receiving the cash, Boruch is alleged to have laundered it by converting it into WebMoney for his fraudster clients, who sent it to their co-conspirator overseas. A trial for Boruch is pending. (At the time of publication, Boruch stood convicted of conspiring to evade currency reporting requirements.)

We also worked cooperatively with the Dominican National Police to dismantle a ring of identity thieves operating from the Dominican Republic. Participants in this transnational conspiracy were filing thousands of fraudulent federal income tax refund claims using identities stolen from residents of Puerto Rico. And they were filing all of these returns with the United States taxation authority electronically, from computers in homes and apartments scattered throughout an area of the Dominican Republic, believing that they were safe from law enforcement. Together with our partners in the Dominican Republic, during the first few weeks of 2009 we tracked these criminals over the Internet by their IP addresses as they filed over 8,000 tax returns seeking over $90 million worth of fraudulent refunds. In late February 2009, together with Dominican law enforcement, we conducted simultaneous arrests of a total of fourteen individuals in New York City and in the Dominican Republic.

These are just a few examples of the cyber crime cases that the Southern District of New York has been investigating. Together, they show that collaboration is vital, prosecution is possible, and deterrence is achievable.

Private Sector Collaboration

As I mentioned previously, cyber crime cases require collaboration, not only among law enforcement agencies, but also between government and industry. Victim companies are on the front lines of this battle, and are often the first to realize that a cyber attack has taken place. Unless we know about the problem, our ability to help is limited. That is why our private sector partnerships are so important.

We understand that there has traditionally been a dichotomy between businesses wishing to maintain the security of their networks and law enforcement investigations. Businesses may feel that they lose control of the process, and that confidential business information could be exposed. Companies might even think that reporting a breach may harm their competitive advantage.

But businesses should know that we will do everything we can to minimize disruptions to their operations, and to safeguard confidential data. Where necessary, we will seek protective orders—orders signed by a federal judge that help to preserve trade secrets and business confidentiality. And we will share with victim companies what we can, as fast as we can, about a particular attack.

The private sector, particularly international companies, is on the front lines of this fight. Many of them have know-your-customer policies and anti-fraud systems, which not only keep out unwanted intruders and prevent criminals from making use of legitimate business channels, but give advance warning of where the next cyber crime wave will come from.

At the end of the day, we can and must help each other by maintaining clear lines of communication between law enforcement and the businesses that are affected by cyber crime.

Conclusion

Cyber crime respects no boundaries. It presents a grave threat to this country, and its perpetrators are constantly honing their attacks. No one country, no one agency, and no one company can alone stop cyber attacks. It is only together that we can minimize this law enforcement and national security challenge. We have had great success prosecuting cyber criminals so far, and I know that by working together we will continue and will expand on this success.

Given what is at stake for our economy and our security, we have no choice.

Cyber Security: Safeguarding Our Cyberspace

Robert S. Mueller III

It is perhaps a little unusual to start a speech by pausing for five seconds, but that is what I would like to do.

(PAUSE)

What just happened? In those five seconds, computer users conducted some 170,000 Google searches. An estimated 22 million e-mails were sent—and about 80 percent of those were spam. Users posted at least 3,500 status updates on Facebook and 3,000 "tweets" on Twitter. Meanwhile, the Automated Clearinghouse—the network that connects all U.S. financial institutions—processed almost 3,000 electronic payments. All of that happened in just five seconds.

We live in a wired world. Our networks help us to stay in touch with family and friends, collaborate with colleagues worldwide, and shop for everything from books to houses. They help us manage our finances and make businesses and government more efficient. But our reliance on these networks also makes us vulnerable. Criminals can use the Internet to commit fraud and theft on a grand scale, and to prey on our children. Spies and terrorists can exploit our networks to steal our secrets, attack our critical infrastructure, and threaten our national security. And because the web offers near-total anonymity, it is difficult to discern the identity, the motives, and the location of an intruder. Yet for too many individuals and businesses, cyber crime remains a nebulous concept. In this chapter, I discuss the evolving nature of cyber threats, what the FBI is doing to combat them, and how we can work together to keep them at bay.

Cyber Terrorism

Let me begin with cyber threats to our national security. As you well know, a cyber attack could have the same impact as a well-placed bomb. To date, terrorists have not used the Internet to launch a full-scale cyber attack. But they have executed numerous denial-of-service attacks and defaced numerous websites. In the past decade, al-Qaeda's online presence has become almost as potent as its physical presence. Extremists are not limiting their use of the Internet to recruitment or radicalization; they are using it to incite terrorism.

Of course, the Internet is not only used to plan and execute attacks; it is also a target itself. Osama bin Laden long ago identified cyberspace as a means to damage both our economy and our morale—and countless extremists have taken this to heart.

We in the FBI, with our partners in the intelligence community, believe the cyber terrorism threat is real, and is rapidly expanding. Terrorists have shown a clear interest in pursuing hacking skills. And they will either train their own recruits or hire outsiders, with an eye toward coupling physical attacks with cyber attacks.

Apart from the terrorist threat, nation-states may use the Internet as a means of attack for political ends. Consider what took place in Estonia in 2007 and in the Republic of Georgia in 2008. Wave after wave of data requests shut down banks and emergency phone lines, gas stations, and grocery stores, and even parts of each country's government. The impact of these attacks left all of us aware of our vulnerabilities.

Counterintelligence and Economic Espionage

Let me turn for a moment to counterintelligence intrusions and economic espionage. Espionage once pitted spy versus spy and country against country. Today, our adversaries sit on fiber-optic cables and Wi-Fi networks, often unknown and undetected. They may be nation-state actors or mercenaries for hire, rogue hackers or transnational criminal syndicates. These hackers actively target our government and corporate networks. They seek our technology, our intelligence, our intellectual property, and even our military weapons and strategies. In short, they have everything to gain, and we have a great deal to lose. We are concerned not only about the loss of data, but corruption of that data as well. If hackers made subtle, undetected changes to your company's source code, they would have a permanent window into everything you do. Some in the industry have likened this to "death by a thousand cuts." We are bleeding data, intellectual property, information, and source code—bit by bit, and in some cases, terabyte by terabyte.

The solution does not rest solely with better ways to detect and block intrusion attempts. We are playing the cyber equivalent of cat and mouse, and, unfortunately, the mouse seems to be one step ahead. We must work to find those responsible. And we must make the cost of doing business more than they are willing to bear.

The FBI: Protecting Our Infrastructure

The FBI pursues cyber threats from start to finish. We have cyber squads in each of our fifty-six field offices around the country, with more than 1,000 specially trained agents, analysts, and digital forensic examiners.

Together, they run complex undercover operations and examine digital evidence. They share information with our law enforcement and intelligence partners. And they teach their counterparts—both at home and abroad—how best to investigate cyber threats. But the FBI cannot do it alone.

The National Cyber Investigative Joint Task Force includes eighteen law enforcement and intelligence agencies, working side by side to identify key players and schemes. The goal is to predict and prevent that which is on the horizon, and to pursue the enterprises behind these attacks. The task force operates through Threat Focus Cells—smaller groups of agents, officers, and analysts from different agencies, focused on particular threats.

For example, the Botnet Focus Cell investigates high-priority botnets. We are reverse-engineering those botnets, with an eye toward disrupting them. And we are following the money wherever it leads, to find and stop the botmasters. The recent takedown of the Mariposa botnet is but one example of that collaboration. As you may know, Mariposa was an information-stealing botnet—one that infected millions of computers worldwide, from Fortune 500 companies to major banks. During a two-year investigation, the FBI worked closely with our overseas counterparts to track down and arrest the main operators of the Mariposa botnet and the original creator of the malicious software that helped to build and control it. In February, the Spanish police arrested three individuals who used Mariposa to hack into online bank accounts. And just two weeks ago, the Slovenian police identified and arrested the botnet's creator. This individual had sold the original virus to hundreds of criminals worldwide, and developed customized versions to meet their needs.

The Mariposa takedown sends a clear message to cyber criminals: We are going after both the cyber equivalent of the house burglar, and the person who gives him the crowbar, the map, and the locations of the best houses in the neighborhood.

The skill, dedication, and unprecedented cooperation provided by our partners in Spain and Slovenia were crucial to the success of this effort. In international cases such as this, global cooperation is absolutely essential.

To that end, the FBI has sixty-one legal attaché offices around the world, sharing information and coordinating investigations with our host countries. We have embedded agents with police forces in Romania, Estonia, Ukraine, and the Netherlands, to mention just a few. Together, we are making progress. But law enforcement agencies alone cannot defeat our cyber adversaries. In the Mariposa case, our private sector partners also provided valuable help. The Mariposa Working Group, an informal band of security researchers and volunteers, gave us intelligence to track down the subjects, and worked to dismantle the botnet after we made our arrests.

Importance of Private Sector Partnerships

To stem the rising tide of cyber crime and terrorism, we also need the help of the private sector.

We in the FBI understand that those in the private sector have practical concerns about reporting breaches of network security. They may believe that notifying the authorities will harm their competitive position. They may have privacy concerns. Or they may think that the information flows just one way—and that is to us. We do not want them to feel victimized a second time by an investigation. We will minimize the disruption to their business, and safeguard their privacy and data. Where necessary, we will seek protective orders to preserve trade secrets and business confidentiality. And we will share with them what we can, as quickly as we can, about the means and the methods of attack.

Remember that for every investigation in the news, there are hundreds that will never make the headlines. Disclosure is the exception, and not the rule. That said, we cannot act if we are not aware of the problem. Maintaining a code of silence will not benefit anyone in the long run.

It calls to mind the old joke about two hikers in the forest who run into a bear. The first hiker says to the other, "We just need to outrun him." And the second replies, "I don't need to outrun him. I just need to outrun you." You may well outrun one attack, but you aren't likely to avoid the second, or the third. Our safety lies in protecting not just our own interests, but our critical infrastructure as a whole.

Conclusion

Following World War I, France built a line of concrete fortifications and machine gun nests along its borders. It was designed to give the French army time to mobilize in the event of an attack by Germany. The secondary motivation was to entice Germany to attack Belgium as the easier target.

As we all know, the Maginot Line held strong for a brief time. However, in the long run, it failed. The Germans invaded Belgium, outflanked the line, and stormed France. In the end, neither fortresses nor fortifications stopped Nazi Germany.

Our success in defeating Germany was built on a united front. We stopped playing defense, and we pushed back, day by day. No one country, standing alone, could have ended that war.

The same is true today, in this new context. No one country, no one company, and no one agency can stop cyber crime. A "bar the windows and bolt the doors" mentality will not ensure our collective safety. Fortresses will not hold forever; walls will one day fall down. We must start at the source; we must find those responsible.

The only way to do that is by standing together. For ultimately, we all face the same threat. Together, we can and we will find better ways to safeguard our systems, minimize these attacks, and stop those who would do us harm.

Cyber Security: Securing Our Cyber Ecosystem

Howard A. Schmidt

More than a decade ago, cyber crime was seen as a high school "hacker" trying to break into a system to prove his or her computer savvy. What were once regarded as the simple pranks of clever minds have evolved into well-organized criminal activity threatening both world commerce and the safety and security of a country's infrastructure.

In 2010, the Internet served as a trading platform for $10 trillion in business. This number will more than double in ten years. Yet many small businesses, a little less than 50 percent of them, do not use antivirus software and even fewer of them who have such software update it on a regular basis.

Because many operations do not strongly focus on securing their information, the Internet is ripe for corporate espionage and modern-day buccaneers whose sole purpose is to infiltrate network systems bent on economic havoc. Who are these people? They operate through many links and they are almost impossible to trace. They can be sitting next to you or sitting in a room on another continent 10,000 miles away. Yet their acts are clearly malicious.

Financial gain from creating malware is a common motive, so much so that criminals even provide technical support for their products. There has been enough exfiltration of personal property in this country in the past years to fill the Library of Congress over and over again. So we all must do more to protect our systems.

Herein lies the heart of cyber security: security is a shared responsibility. We are all stakeholders—federal, state, and local governments, law enforcement agencies, private corporations, academia, and individual citizens. Public-private partnerships must find ways to create a technology road map to prevent the misuse and abuse of information. We must create the vision and guidelines for sharing confidential information safely. And while we work toward a solution, we must also ratchet up the fight against cyber crime.

To foster greater responsibility for the protection and safety of information, the work being done in the academic and private sectors is of great interest to the federal government. The federal government is focused on encouraging a trusted identity system. I want to add that as part of this work, in 2010, the White House established the National Strategy for Trusted Identities in Cyberspace (NSTIC). The NSTIC is a response to a near-term action item in President Obama's Cyberspace Policy Review. The strategy calls for the creation of an Identity Ecosystem. The ecosystem's core is for the key components of a cyber transaction, namely the individual and organization identities, along with the identities of the infrastructure that handles transactions, to operate in a streamlined and safe manner, moving away from the culture of having different user names and passwords—which are often the same or never changed—for each website. In its place, individuals voluntarily choose a secure privacy-enhancing credential to verify themselves for all types of online transactions from online banking, sending email, maintaining health records, or for any other personal cyber uses. To achieve such a streamlined operation, however, security, efficiency, ease-of-use, confidence, privacy, choice, and continued opportunities for innovation must be the Identity Ecosystem characteristics. Privacy protection and voluntary participation are the fundamental building blocks needed to achieve Identity Ecosystem success. This is the road before us, and I welcome your participation on this initiative.

Of course, for such a project to leap forward, there is the obvious need for complete confidence in the security of both personal and business information. This is where the Comprehensive National Cyberspace Initiative (CNCI) plays the fundamental role in addressing cyber criminal activity. The CNCI outlines a plan for sharing situational awareness among federal, state, and local governments, and private industry partners. Other initiatives are the ability to react quickly to threats and incursions, the ability to defend against cyber threats through enhanced counterintelligence, protecting the safety of information technology pathways, and strengthening the cyber security environment through expanded cyber education.

Research and development are the crucial elements in assuring these safeguards. Development of resilient systems in networks for better backups will help mitigate disruptions, and development of moving target technology that floats critical information within a network will help protect information by making it harder for malicious actors to locate.

One action, though, is of paramount importance for the safety and security of the United States: identifying and securing Internet connection points to government networks. While the Internet grew rapidly, it has done so in a wild and free manner, like a spiraling tornado of information. It is clear that after the incredible two-decades-long expansion of Internet usage, a cleanup of government computer networks is warranted. A decisive movement that limits connections to the Internet, called Trusted Internet Connections (TIC), will increase law enforcement capabilities and allow for secure sharing of information across government agencies.

At the same time, we can't lose sight of privacy. The core principle that privacy and civil liberties will not take a backseat to better security is part of who we are.

This is why protecting information comes down to protecting yourself. Maybe it's just something you do differently. Just changing your password helps us develop better trust in cyberspace. There also has to be cognitive awareness that information never dies on the Internet. Whatever is entered today will likely be there generations from now, floating somewhere, and with some effort, retrievable. The key is that we each have to do our part to secure our part of cyberspace.

HIRA AGRAWAL, PHD is a senior scientist at Applied Communication Sciences (formerly Telcordia Technologies). He has over twenty years of research experience in software engineering techniques and tools. He is currently leading a U.S. Army CERDEC project on software quality assurance. He received his PhD in computer science from Purdue University.

PREET BHARARA, JD is the United States attorney for the Southern District of New York. He oversees the litigation of all criminal and civil cases brought on behalf of the United States in the district. A graduate of Harvard College, he received his JD degree from Columbia Law School where he was a member of the *Columbia Law Review*.

THOMAS F. BOWEN, PHD is a chief scientist in Applied Communication Sciences (formerly Telcordia Technologies) Applied Research Area, performing computer security research focusing on real-time, host-based intrusion detection and reaction for the U.S. government and commercial sponsors. Bowen's prior work with Bell Labs and Telcordia focused on software for control of commercial and prototype telecommunication systems.

DYANN BRADBURY is the president of the InfraGard National Members Alliance (INMA). She previously served on the INMA board where she was chair of the budget committee and a member of the DHS Partnership Committee. She also is the director of IT compliance at Digital River.

VINCENT BUSKENS, PHD is associate professor at the Interuniversity Center for Social Science Theory and Methodology (ICS) at Utrecht University as well as professor of empirical legal studies at the Erasmus School of Law of the Erasmus University Rotterdam.

D. FRANK HSU, PHD is the Clavius Distinguished Professor of Science, former chairman of computer and information science, and former associate dean of the Graduate School of Arts and Sciences at Fordham University. He was chair of the Section of Computer and Information Science at the New York Academy of Sciences and founding editor and editor in chief of the Journal of Interconnection Networks (JOIN). He holds an MS from the University of Texas at El Paso and a PhD from the University of Michigan at Ann Arbor.

KUAN-TSAE HUANG, PHD is the chairman of Taskco Corporation. Prior to that, Dr. Huang was a vice president at IBM and senior consultant at the National Library of Medicine/National Institute of Health. He holds a PhD in EECS from MIT and an MS in mathematics from the University of Illinois, Urbana-Champaign.

KEVIN KELLY, MA retired from the New York City Police Department (NYPD) in 2007 as a detective squad supervisor and executive officer of the Computer Crimes Squad with over 21 years of experience in law enforcement. He investigated and supervised all major criminal investigations where electronics or high technology means were deployed. As the direct supervisor of the computer forensic laboratory, he was responsible for conducting computer forensic investigations, training, equipping, and managing all computer forensic cases in the NYPD. Kelly also helped to develop and teach courses in cyber security and computer forensics at Fordham University.

KATHLEEN L. KIERNAN is a veteran of federal law enforcement and is chairperson of the InfraGard National Members Alliance (INMA). She is also the CEO of Kiernan Group Holdings, an international consulting firm that supports federal and civil clients. She previously served as the assistant director for the Office of Strategic Intelligence and Information for the ATF.

RUBY B. LEE, PHD is the Forrest G. Hamrick Professor of Engineering and a professor of electrical engineering at Princeton University, with an affiliated appointment in computer science. She is director of the Princeton Architecture Laboratory for Multimedia and Security (PALMS). She has a PhD in electrical engineering and an MS in computer science, both from Stanford University, and an AB with distinction from Cornell University. She has been granted more than 120 U.S. and international patents and has authored numerous conference papers, winning several awards.

ANDREW LEWMAN is the executive director of the Tor Project, a nonprofit organization providing research and free software that protects online privacy and anonymity. He has worked for multinational companies in Japan, Korea, Taiwan, and the Philippines.

NICHOLAS J. MANKOVICH, MS, PHD, CIPP is the chief information security officer for Royal Philips Electronics. Formerly he led two worldwide programs for Philips Healthcare Product Security and Privacy. He has held numerous health care and academic appointments including at the University of Iowa, the UCLA School of Medicine, and the University of New South Wales in Sydney, Australia.

DOROTHY MARINUCCI holds an MA in international political economy and development from Fordham University. She serves as the logistics chair for the International Conference on Cyber Security. She is the chief administrator in the Office of the President at Fordham University.

EILEEN MONSMA, MSC is embedded with the preeminent National High Tech Crime Unit (NHTCU) of the Netherlands' Police Agency. The NHTCU aims to investigate and prevent organized high tech crime and primarily works on international cases while applying innovative strategies. The research project in this book is in line with this innovative approach.

ROBERT S. MUELLER III, JD is the sixth director of the Federal Bureau of Investigation, a position he was appointed to on September 4, 2001. He graduated from Princeton University and holds a master's degree in international relations from New York University, a law degree from the University of Virginia, and is a member of the United States Marine Corps.

SANJAI NARAIN, PHD is a senior research scientist in the Information Assurance and Security Department at Applied Communication Sciences (formerly Telcordia Technologies). He received his PhD in computer science from UCLA and an MS in computer science from Syracuse University. His current research is on a science of configuration.

PAUL NIEUWBEERTA, PHD is professor of criminology at the Institute for Criminal Law & Criminology of Leiden University as well as at the Sociology Department of Utrecht University, where his commitment is the theoretical and empirical analysis of the effects of criminal interventions.

ADAM PALMER, JD is the Norton Lead Cyber Security Advisor at Symantec. His work focuses on supporting law enforcement and industry efforts to reduce cyber crime. A former U.S. Navy JAG prosecutor, he has been honored for his work in teaching cyber crime prosecution.

HOWARD A. SCHMIDT is the special assistant to the president and the cyber-security coordinator for the federal government.

MELVIN SOUDIJN, PHD is senior researcher at the Netherlands' Police Agency. This division of the Dutch police supports regional police forces and carries out (inter)national and specialist police tasks.

EDWARD M. STROZ is founder and co-president of Stroz Friedberg, a leading global consulting firm for managing digital risk and uncovering digital evidence. A graduate of Fordham University, Stroz is a certified public accountant, a certified information technology professional and a licensed private investigator.

AKIO SUGENO is vice president of internet engineering, operations and business development at Telehouse America. He is founder of one of the world's largest IXPs: NYIIX and LAIIX. In his current role, he focuses on global business development and strategies. He is actively involved with various engineering organizations such as the Réseaux IP Européens (RIPE) and North American Network Operators' Group (NANOG). He holds a BS from Waseda University.

PAUL SYVERSON, PHD has for twenty-two years conducted research in security and privacy at NRL, where he and his colleagues invented onion routing and designed Tor. He is a graduate of Cornell University and also holds an MA in mathematics and an MA and PhD in philosophy (logic) from Indiana University.

KAZUO TAKARAGI, PHD is the senior chief researcher for the Systems Development Laboratory at Hitachi, Ltd. He has been engaged in the research of system reliability, safety evaluation, cryptography, and information security. He holds a doctorate degree in safety engineering from the University of Tokyo.

HWAI-JAN WU, PHD is the vice president of Taskco Corporation. Prior to joining Taskco, she was an associate professor at the Department of Management Information System at Chung-Yuan Christian University, Taiwan. She holds a PhD in industry engineering from Auburn University.